D1524528

CORNBELT REBELLION

The University of Illinois Press, Urbana and London, 1965

CORNBELT REBELLION

The Farmers' Holiday Association

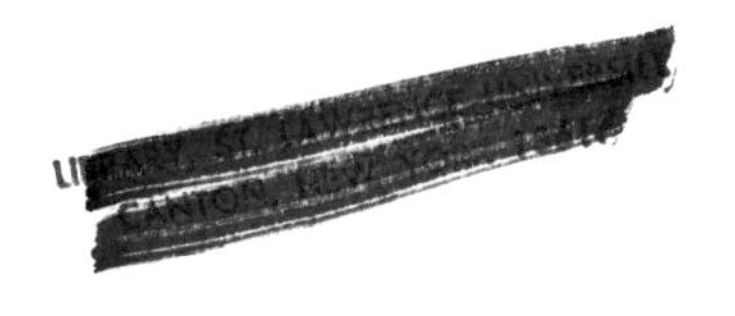

JOHN L. SHOVER

 Manufactured in the United States of America. Library of Congress Catalog Card No. 65-11738.

ACKNOWLEDGMENTS

While I alone must bear responsibility for any errors of omission or commission that follow, this study of the Farmers' Holiday Association has been made possible in large measure by numerous individuals, named and unnamed, who have given generously of their time, their talents, their memories, and personal collections of relevant materials.

My first obligation is to a former student at San Francisco State College, Mr. John Kriege, who more than five years ago brought to me a dusty folder of papers, those of his father, the organizer of the Farmers' Holiday in northeastern Nebraska. That stimulus led to this study. I am indebted to an unknown mail clerk in Des Moines, Iowa who misdirected a letter inquiring of the possible existence of papers of Milo Reno into the hands of Mr. Fred Stover, president of the U.S. Farmers' Association. Mr. Stover not only turned over to me for personal use the invaluable file from the back closet of his office consisting of the papers of the Farmers' Holiday Association for the year 1933, he also drew upon his wide knowledge of farm organization politics to provide counsel and insight and he made contact for me with a score of important individuals, many of whose names appear in the bibliography under the caption "Personal Interviews." They were veterans of the Farmers' Holiday Association.

Mr. Lem Harris of New York City admitted me, a total stranger, into his confidence and friendship, allowed me free access to the files of the Farmers' National Committee for Action which he had carefully preserved for thirty years, and patiently abided our disagreements in many hours of discussion and several dozen letters in order to assist me in presenting the part of the Communist party in the rural unrest of the nineteen thirties.

To those Iowa and Nebraska farmers who joined me in wading through six-foot snowdrifts down farm lanes in Madison County, Nebraska or who provided bed and board while we spent long hours of wintry nights recalling and reconstructing events thirty years past, I can only hope that the account that follows represents some small measure of thanks.

Neither can I express sufficiently my gratitude to the countless intrepid and unhonored librarians who assisted me at the San Francisco State College Library, the University of California (Berkeley), the Nebraska State Historical Society, the State University of Iowa, the Sioux City Public Library, and the Minnesota Historical Society. I must single out for especial mention and thanks Miss Joyce Shober of San Francisco State College and Miss Kathryn Johnson and Mr. Tom Deahl of the Minnesota Historical Society.

Three former colleagues of mine, Professors W. Patrick Strauss, Joel Silbey, and James Flink, have encouraged my study and brought to bear their not inconsiderable critical abilities to challenge my assumptions and ideas. Likewise, Dr. Lowell Dyson, author of a doctoral dissertation on the Farmers' Holiday Association, has generously shared research materials and given me the advantage of testing my ideas against those of a scholar knowledgeable and astute in the identical subject area.

I am grateful to the editors of *Nebraska History, The Journal of American History,* and *Agricultural History* for permission to include here portions of chapters 3, 4, 5, 6, and 8 that have appeared previously in their respective journals.

The American Philosophical Society, through a generous grant from their Penrose Fund, made possible a research trip through Iowa, Nebraska, and Minnesota in the spring of 1962.

Lastly, to my wife, Ann, I owe a special obligation beyond her typing and editorial assistance, for when she married me four years ago little did she know she was marrying the Farmers' Holiday Association as well. Fortunately, despite the magnitude of the shock, she chose not to foreclose the otherwise happy arrangement.

Mill Valley, California
December, 1964

CONTENTS

INTRODUCTION

The cyclical economic recessions recurring in America since the Civil War have produced a crop of agrarian protest movements that have sprung up abruptly, flourished with brief intensity, and disappeared as rapidly as they arose. Greenbackers, Populists, or the Non-Partisan League have won strong minority support from farmers and in times of economic crisis rural America has abandoned its traditional political association with the party of high tariffs, Eastern capital, and limited government.

Basic to all these rural protests was the fact that the farmer, unlike other small businessmen, offered his produce in a buyer's, not a seller's market. The farmer was a speculator, taking his chance each year on a profitable return at harvest time. When these returns sank so low that credit obligations could not be met, the ordinarily apathetic farmer was ready to turn to voluntary associations, to government, and, in a few instances, to direct action in order to right the situation.

The American family farm remained an isolated outpost of traditional free enterprise for long after the economic system passed it by. A common denominator in the panaceas rural American has produced, from sub-treasury and free silver to state-administered crop insurance and state-owned elevators, has been the attempt to catch step with an economy characterized by rational price and production controls. Farmers through their protests won a series of transitory victories from the agents of corporate capitalism: fairer railroad rates, protection for cooperative marketing arrangements and even, in the Agricultural Adjustment Administration, a method of imitating production controls. But in the long run the family farmer has not persevered against the system itself. Technological improvements reducing the ratio of labor to output, heavy capital costs, and the advantages of mass production have brought to agriculture the same consolidation movement which long ago smothered the corner grocer and the local craftsman.

The Farmers' Holiday was the most aggressive agrarian upheaval of the twentieth century. Spanning the transitional period of the early thirties, it was a final great attempt of the family farmer to save himself from absorption and annihilation. The resistance has not ended. As these words are written another farm movement, the National Farmers' Organization, with objectives and methods remarkably similar to the Holiday Association, is gathering followers in the Midwest. Possibilities for its success are not great. In 1962 only 10 to 15 per cent of total agricultural products originated from small farms;[1] corporation farming expanded rapidly; young farmers were uncommon in the rural population. The weight of economic opinion criticizes the small, diversified farm as an inefficient and uneconomical operation. The program of this contemporary protest organization, like the Holiday, is economic but the ultimate rationale is ideological: the defense of traditional rural and small town culture.

The actions and program of the Holiday Association, as of earlier protests, were often simplistic, aimed merely at the immediate remedy of adverse economic conditions. However, to dismiss these protests as colorful aberrations of unsophisticated minds misses the point. Milo Reno, despite his evangelism and demagoguery, realized what was at stake. Misconceived farm strikes or brave attempts to save a brother farmer from foreclosure were symptoms of broad social and economic changes made more manifest by depression conditions.

The agricultural program of the New Deal quelled the unrest in the Midwest, but like the farmers' own protests, it too failed to check the steady trend toward consolidation. The very title of the agrarian program bespeaks that it was an "adjustment" policy only; the government's benefit payments served finally to cushion and assist the large operator more than the family farmer.

The cornbelt rebellion of the thirties is a final chapter in a story that begins with Daniel Shays and ends with Milo Reno.

---

[1] Jackson V. McElveen, "Farm Number, Farm Size and Farm Income," *Journal of Farm Economics,* XLV (February, 1963), 11.

One

AN AUDIENCE FOR THE AGITATORS

There are still plenty of farmers who have rich land, unmortgaged, work it intelligently, and even at the lowest prices are making a pretty good living. They are not joining any movement of protest. There are others who have been put completely down and out, like the South Dakotans who have suffered from drouth, grasshoppers and low prices so long and so completely that they are flat on their backs. They don't do any hollering. It's where the farmers had something a few years ago and have had it suddenly taken away, that the agitators find a responsive audience [Bruce Bliven, "Milo Reno and His Farmers," *New Republic,* LXXVII (November 29, 1933), 64].

"Stop! Farmers' Holiday!" read the sign along the roadway. A cluster of uneasy men in overalls waited as the farm trucks crept up hill in low gear. A log and a threshing machine belt were held ready to throw in the path of the oncoming vehicles. As the trucks ground to a halt there was quiet talk, sometimes words of anger, but usually the trucks turned back. By the night of August 14, 1932, some 1,500 farmer pickets, guarding all the roads to the city, had virtually halted all milk and livestock deliveries to the terminal market at Sioux City, Iowa. The hogs the trucks carried would have brought the farmer that day just 3¢ a pound, the milk, 2¢ a quart.[1]

This was the American cornbelt, one of the world's most favored agricultural areas. Here nature had been beneficent; rainfall was regular and calamities such as drouth and insect

[1] *Sioux City* (Iowa) *Journal,* August 15, 1932; A. G. Black, "The Situation Today," in A. G. Black *et al, The Agricultural Emergency in Iowa,* Circular 140 (Ames, 1933), 1-2.

plagues less common than in wheat or cotton producing regions. Livestock, the principal product of the area, was less affected by the vagrant price and weather fluctuations that troubled less diversified wheat growers to the north and west.[2] In Iowa, most typical of cornbelt states, an average rural family lived on a 150-acre farm in a house both larger and better equipped than farm homes in any nearby state. Half of the families owned a radio and an Iowa farmer was more likely to have a telephone and an automobile than a farmer in any neighboring state.[3]

The farm strike centered in some of the most prosperous farming country in the cornbelt. Of the nine Iowa counties that accounted for more than three-quarters of the incidents of rural protest in August and September of 1932, seven were located at the western fringe of the state adjacent to the Missouri River. In this, the leading meat production sector of the state, gross income per farm was well above the state average. Few other clear distinctions separate the nine counties with highest protest intensity from the other eighty-nine of Iowa's counties. The percentage of tenancy in each high intensity county was above the state average of 50 per cent, but there were counties in north-central Iowa (with no protest activity) where 75 per cent of farms were cultivated by tenants. In both land value and mortgage debt per acre the high intensity counties ranked in the middle range of Iowa counties, neither the highest nor the lowest. Nativity composition was diverse, but no more so than most Iowa counties.[4] In one ecological area, however, the farm strike excited

---

[2] Corn was customarily fed to livestock, not marketed. Iowa led the nation in hog production: in 1930, 42% of the state's farm income was derived from this source. Lauren K. Soth, *Agricultural Economic Facts Basebook of Iowa* (Ames, 1936), 9, 38.

[3] Charles P. Loomis and J. Allan Beegle, *Rural Social Systems* (New York, 1950), 262-66; Soth, 152-59.

[4] To determine counties with most intense protest activity a tabulation was made of reports of protest activities during August and September, 1932 from two leading area newspapers, the *Des Moines Register* and the *Omaha World-Herald*. Each occurrence was classified by county; an event reported in both newspapers was recorded twice, avoiding purely local reporting and giving double weight to an incident important enough to have been mentioned in both journals. The nine counties that scored highest were in order: Woodbury, Plymouth, Pottawattamie, Polk, Harrison, Monona, Cherokee, Clay, and Black Hawk. Three hundred and thirty-five incidents of protest were recorded, 77.6% in these nine counties. Conclusions as to economics and demography are based upon data assembled in Soth, 10, 104-5, 114, 117, and 139 and in William L. Harter and R. E. Stewart, *The Population of Iowa: Its Composition and Changes*, Bulletin 275 (Ames, 1930), 27-31.

Iowa Counties of Important Protest Activity, 1932-33

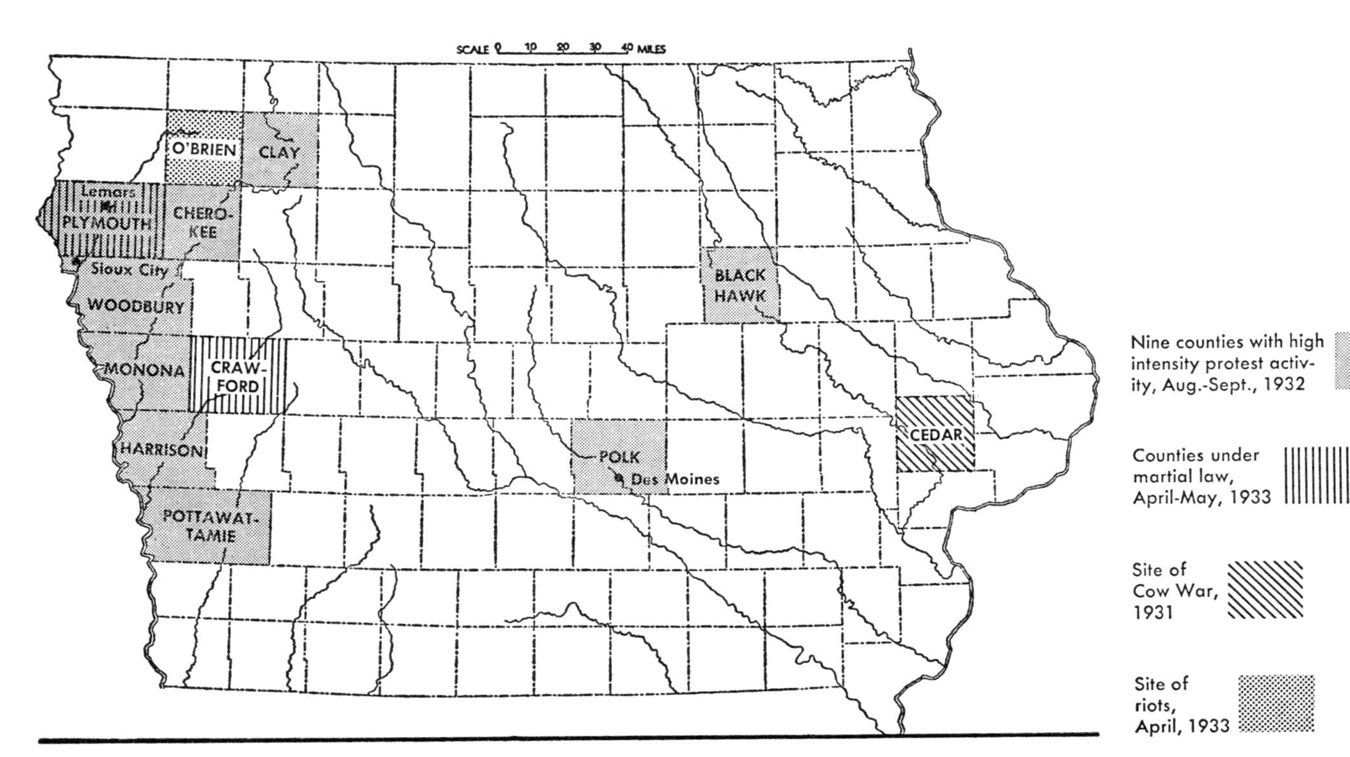

little interest: this was the tier of counties in the so-called southern pasture area touching the Missouri border. Here land value per acre and gross income per farm were the lowest in the state, tenancy was the highest, and home conveniences were fewest. This, the poorest economic area in Iowa, was least responsive to Farmers' Holiday agitation.[5]

Apart from the general factor of higher gross income per farm, two variables immediate and critical in nature separate the high intensity counties in the west from other Iowa counties.

First was the weather. Iowa rainfall in 1931 was 10 per cent above normal with one exception—the counties in the northwest corner including Woodbury, Plymouth, and Cherokee where strike activity was most spirited a year later. Deficiencies in 1931 ranged from 6 to 8 inches below the norm and the corn crop was depleted several bushels per acre. Serious drouth again threatened in 1932 as twenty-eight days of the prime growing season in June and July passed without rainfall in Monona, Woodbury, and Plymouth Counties. Copious rain in August, widely distributed throughout the state, remedied the situation and Iowa produced in 1932 one of the largest corn crops in its history. Only one small area reported below average precipitation in August. This was a pocket in Woodbury and Plymouth Counties—the counties that rank first and second respectively in the incidence of farm strike activity.[6]

The second variable was the number of foreclosure suits. Statistics have not been gathered for all Iowa counties, but this much is certain: *Woodbury (Sioux City), the county with the greatest strike activity, had three times more foreclosure suits pending (485) than any other Iowa county.* Two other of the nine high intensity counties, Black Hawk and Plymouth, ranked among the eleven counties with most suits pending.[7]

---

[5] Soth, 160. A similar circumstance can be observed in other protest movements. For example, the earliest support for the Cooperative Commonwealth Federation of Saskatchewan came from the upper layer of farmers and workers groups; the more depressed segment grew interested in periods of rising prosperity. S. M. Lipset, *Agrarian Socialism* (Berkeley, 1950), 175-76.

[6] U.S., Department of Agriculture, Weather Bureau, *Climatological Data, Iowa Section,* XLII, No. 13 (Annual, 1931), 99-100; XLIII, No. 7 (July, 1932), 54; XLIII, No. 8 (August, 1932), 66.

[7] *Des Moines Register,* July 15, 1934; see also William G. Murray and Ronald C. Bentley, "Farm Mortgage Foreclosures," in A. G. Black *et al,*

In sum, protest in Iowa emanated from relatively prosperous areas where some immediate crisis, i.e., drouth or foreclosure, threatened to deprive farmers of property or accustomed income. "It's where the farmers had something a few years ago and have had it suddenly taken away, that the agitators find a responsive audience," wrote Bruce Bliven, a native Iowan.[8]

With little exception the counties where strike activity was most manifest had paralleled the consistent voting pattern in Iowa.[9] Since 1900 the state had never wavered in allegiance to the Republican party. Nevertheless, as in other farm states, successful Republican candidates were often political independents who identified in name only with the national program of the party. Dan W. Turner, elected governor in 1928, was a progressive in the George Norris tradition. Two times in the twenties Iowa sent to the Senate Smith W. Brookhart, a political maverick who bitterly assailed the vetoes of the McNary-Haugen bill and Republican inaction in agriculture. Analysis of the votes for Brookhart suggests that northwest Iowa was politically restless territory. When Brookhart first campaigned in 1920 he received little assistance from these counties, but in his second campaign in 1924, the greatest percentage increase in support was in the northwest, confronted at the time with a serious drouth. A student of the Brookhart campaigns concludes, "The meat producing areas inclined toward Brookhart most strongly when prices of livestock and livestock products were relatively poorest." [10]

Insurgent Republicanism aside, rural protest movements of the past had made little headway in Iowa or in the counties where the cornbelt rebellion of the thirties centered. In 1892 the state returned only 5 per cent of its votes for its native son, the Populist candidate for president, James B. Weaver. The Greenback party, Populism, and the Non-Partisan League all began in wheat pro-

---

The Agricultural Emergency in Iowa, Circular 147 (Ames, 1933), 177 and Lawrence A. Jones and David Durand, *Mortgage Lending Experience in Agriculture* (Princeton, 1954), 88.

[8] "Milo Reno and His Farmers," *New Republic,* LXXVII (November 29, 1933), 64.

[9] The single exception was Plymouth County's vote for Smith in 1928. Edgar Eugene Robinson, *The Presidential Vote, 1896-1932* (Stanford, 1934), 83-85, 192-200.

[10] Jerry A. Neprash, *The Brookhart Campaigns in Iowa, 1920-1926* (New York, 1932), 30-60, 120.

ducing areas and maintained their greatest strength there. Dairy and livestock farmers, like those in Iowa, showed slight interest. On the basis of this kind of experience, students of rural protest tend almost automatically to equate insurgency with grain growing areas where there is little crop diversity and price fluctuations are extreme. Trade monopolies exercise more power, grading procedures are more complex, and freight costs are higher than in corn and dairy regions.[11] A pioneer study in political behavior, charting the spread of the Non-Partisan League, identified a "relatively conservative area" in the Missouri River counties beginning in the southeast corner of South Dakota and extending south along both banks of the river in Iowa and Nebraska. "It is evident that this prosperous farm region has offered resistance to the diffusion of radical or progressive political opinions that has not been encountered to an equal extent to the north, east or west of it."[12] Defying such a generalization, this very area spawned the violent rural protest of the thirties and always remained the center of activity. The depression farmers' movement was a new phenomenon in agrarian uprisings. It was primarily a movement of corn-hog and dairy farmers. Counties in Iowa where protest was most intense included the leading hog producing counties; areas outside of Iowa where insurgency had the greatest attraction were all pork and milk producing regions.

The characteristics of the individuals who participated in the protest movement are more difficult to determine, but some evidence is available on the relative incidence of property holders, tenants, and farm laborers. Of ninety individuals arrested from the picket lines at Sioux City on August 26, five were property holders; twenty were former owners (now renters); twenty-five

---

[11] Lipset, 10-11; Clyde O. Ruggles, "The Economic Basis of the Greenback Movement in Iowa and Wisconsin," *Proceedings of the Mississippi Valley Historical Association for the Year 1912-1913,* VI (1913), 142-65; Benton H. Wilcox, "An Historical Definition of Northwestern Radicalism," *Mississippi Valley Historical Review,* XXVI (December, 1939), 384-85; Herman C. Nixon, "The Economic Basis of the Populist Movement in Iowa," *Iowa Journal of History and Politics,* XXI (July, 1933), 373 and *passim.* Counties in northwest Iowa had given Populism greater support than other areas of the state, but there was a particular reason for this which is no longer present. In the nineties this area was the last wheat growing frontier in a state rapidly shifting to more stable corn and hog production.

[12] Stuart A. Rice, *Farmers and Workers in American Politics* (New York, 1924), 180.

were tenants; fifteen, farm youths living at home; seventeen, farm laborers; six, city workers and two, unknown.[13] This is the only data that suggests anything other than that the farm protest was made up largely of property holders. Those who picketed may have represented a different population from those who were leaders or even members of the Holiday Association; even in this group those selected for arrest would probably constitute the most belligerent element among the pickets. Given this, the presence of twenty-five owners or former owners seems significant. Most important, both leaders and participants had the image of the farm strike as a movement of farmer landowners. As they looked back upon the movement from the vantage point of thirty years, not one of eighteen individuals who were interviewed by the writer believed the Farmers' Holiday was a movement of tenants and farm laborers. Most believed it consisted largely of men who owned property or "about half and half" tenants and owners.[14] The ideological appeal of the Farmers' Holiday Association was oriented to property holders. "We believe with Oliver Goldsmith," wrote the president, "that every man should own his rood of ground." To a correspondent who challenged this viewpoint, he replied, "I cannot hardly agree with you . . . that the farmers must understand that they are only laborers and not businessmen and capitalists." [15]

"The farmers of the Mississippi Valley are in a desperate way, not only financially, but in the matter of morale," wrote Milo Reno, the leading evangelist of rural protest in Iowa in 1932. "For twelve long, weary years we have pled for the recognition

[13] *Iowa Union Farmer* (Columbus Junction, Iowa), September 7, 1932.

[14] One of these, W. C. Daniel, president of the Woodbury County Farmers' Holiday Association, insisted the newspaper statistics on the number of property holders arrested were wrong. When I read to him the list of names of those arrested from Woodbury County in this and other newspaper sources, he recalled a large percentage of them as owners. A further bit of evidence is a questionnaire circulated to a random group of Holiday veterans by the writer. Of thirty-one farmer respondents, twenty-five had been property holders in 1932-33.

[15] Milo Reno to Mrs. Garrit Jansma, October 18, 1933; to M. W. Hennessy, November 22, 1933. Milo Reno Papers, Library of the State University of Iowa, Iowa City. All Reno letters or manuscripts hereinafter cited are a part of this collection unless otherwise indicated.

that we felt we were entitled to, only to be insulted or ignored." [16] The depression farmers' movement took shape in the twenties. The economic collapse of the early thirties aggravated a crisis and activated a sentiment that had been building in the midwestern countryside for twelve years.

If ever there were halcyon days on the farm they were the war years of 1918 and 1919. American farmers supplied a world market and prices spiraled accordingly. Using the 1910-14 base as 100, farm prices in 1919 stood at an unprecedented 213, more than double 1915 returns. In the optimism of new-found prosperity, land, equipment, and buildings were mortgaged to buy more and more farmland. Land prices rose to 160 per cent of the prewar average.[17]

The fair-weather economic climate disappeared as quickly as it had come. The swollen war market contracted as foreign competitors returned to production; inflated prices dwindled to prewar levels. The one inheritance the farmer carried with him from the bountiful years was an unchanged burden of debt accumulated when times were good and omens favorable.

International exports normally constituted a small portion of total agricultural output, but aggravated increases or decreases in overseas demand could often exercise a determinative influence over farm prices. Until the turn of the century the bulk of total American exports had been agricultural products. These balanced American payments in the international market, making up deficits acquired through imports of manufactured goods and raw materials.[18] The waning of the foreign market after 1900 inaugurated no agricultural recession; rather, the years from 1900 to 1914 were a "golden age" of American agriculture. Industrial production increased rapidly enough to offset increments in agricultural production over the preceding four decades. Increasing urbanization, accompanied by rising wages and employment,

---

[16] Letter to W. P. Meakin, May 18, 1932. Franklin D. Roosevelt Papers (Official Files, Agriculture), Hyde Park, New York.

[17] U.S. Department of Agriculture, *Yearbook of Agriculture, 1935,* 681; H. C. M. Case, "Farm Debt Adjustment During the Early 1930s," *Agricultural History,* XXXIV (October, 1960), 173-74.

[18] For example, between 1877 and 1881, agricultural produce made up 80.4% of total exports. William Trimble, "Historical Aspects of the Surplus Food Production of the United States, 1862-1902," *Annual Report of the American Historical Association for the Year, 1918* (Washington, 1921), I, 227.

meant that a larger share of farm produce was domestically consumed. Consumption of pork, for example, rose 50 per cent. Farmers, in turn, learned to adjust more effectively to market demands. Shifting from sole reliance on staples, there was a trend toward more diversified and profitable products such as fresh milk for cities or out-of-season fruits and vegetables. The results were gratifying for both market prices and land values rose. Between 1900 and 1914 supply and demand for farm products stood in effective equilibrium.[19]

The artificial stimulus of wartime demands dissipated a harmony that had prevailed for nearly two decades. The collapse of the foreign market after 1920 had its most pronounced effect on wheat and hogs, the two products whose volume had expanded most in wartime. Not only was there a return to production in Europe, but output of wheat increased in competing surplus areas like Canada and Argentina. Hog production in Germany and Denmark, the two principal European competitors, reached prewar proportions in 1927. To these competitive factors were added protective restrictions by major foreign importers. By 1933, Germany had increased the lard tariff to almost twice the value of the product at the current exchange rate and in 1932, England, the most important market for pork products, initiated a system of import quotas designed to increase domestic production and prices. The Smoot-Hawley tariff of 1929, choking sources of foreign exchange in the United States, made sale abroad of agricultural surplus a virtual impossibility. Wheat exports declined from a high of 37 per cent of the total crop in 1920 to 17 per cent in 1923, rallied in 1924, then plummeted to 17 per cent in 1929 and 4.3 per cent in 1933. Pork exports in 1919 totaled 24 per cent of total production, in 1926, 6 per cent and in 1932 only 1.7 per cent. On the eve of the eruption in the Midwest, the Secretary of Agriculture reported that total exports of hogs for the year (1931) were the smallest in the century. Corn-hog farmers picketed highways at a time when their foreign market was at an unprecedented low.[20]

---

[19] E. Everett Edwards, "American Agriculture—the First 300 Years," U.S. Department of Agriculture (hereinafter abbreviated U.S.D.A.), *Farmers in a Changing World: Yearbook of Agriculture, 1940,* 241.

[20] Murray R. Benedict, *Farm Policies of the United States, 1790-1950* (New York, 1953), 277; U.S.D.A., *Yearbook of Agriculture, 1932,* 29-30; U.S.D.A., *Yearbook of Agriculture, 1935,* 350; D. A. Fitzgerald, *Livestock Under the A.A.A.* (Washington, 1935), 13.

Had consumer demand for wheat and pork grown in the twenties or had purchasing power markedly risen the market lag could have been accommodated. Since this was not the case, farmers were confronted with a serious problem of diminishing real prices throughout the entire decade. Nevertheless, the market price of no farm commodity fell below the 1910-14 average until 1929. The price recession of the twenties was a fact only in terms of the falling away from abnormal highs of war years and in terms of obligations farmers had acquired which required for satisfaction something more than the income that had prevailed before the war.

If the basis of comparison is real prices, the farmer did not fare as well. For example, purchasing power of Nebraska farm products fell from 103 to 72 between 1920 and 1921 and then recovered to a high of 98 in 1926. It declined again to 65 in 1931 and reached a low of 54 in 1932. The buying power of hogs fell to 62 in 1923 but rose to 102 in 1926 due to decreased production. However, the real price diminished to 62 again in 1931 and plummeted to a low of 41 in 1932.[21]

The crash of 1929 sent prices falling faster and further than any period in the history of American agriculture. The market price of a bushel of wheat sank from $1.03 in 1929 to 67¢ in 1930 and a low of 38¢ in 1932. This was below the price during either the Greenback or Populist periods.[22] Hog prices fell even more precipitately from $11.36 per head in 1931 to $6.14 in 1932 and $4.21 in 1933. This was the lowest price since the nineties.[23]

Price declines, more than any single factor, have a relevance in interpreting and predicting agrarian revolt. Farm insurgency is not a matter of type of farming, it is a response to extreme price and market fluctuations. During the Populist period when wheat farmers turned rebel and corn-hog farmers watched apathetically, wheat prices were fluctuating in a pattern like that of the twenties and thirties, but at the same time, hog prices were actually

[21] Soth, 29; Horace C. Filley, "Effect of Inflation and Deflation upon Nebraska Agriculture, 1914 to 1932" (unpublished Ph.D. dissertation, University of Minnesota, 1934), 12.

[22] The lowest price for wheat in the 1890-1900 decade was 49¢ per bushel in 1894.

[23] U.S.D.A., *Yearbook of Agriculture, 1928,* 670; U.S.D.A., *Yearbook of Agriculture, 1932,* 784; U.S.D.A., *Yearbook of Agriculture, 1935,* 567-68.

increasing. Now that situation was changed. As the accompanying chart indicates, for a decade before the great depression, hog prices vacillated in a manner almost exactly parallel to the usually sporadic wheat price. This was a condition to which cornbelt farmers were but little accustomed and, as circumstances proved, not inclined to accept stoically. The cornbelt rebellion, beginning in 1932, came in the wake of the sharpest yearly decline ever recorded in hog prices.

Price declines were a serious problem in the twenties because of too easy credit and too eager speculation in the war period. On the impetus of war prosperity farm mortgage debt skyrocketed hand in hand with land values to 160 per cent of the prewar average. In Iowa slightly more than half of the farms were mortgaged, a figure not excessive, but inflated land values meant that the farmer debtor carried an unusually heavy burden. The average size of new loans in Iowa was $11,080 in 1920 compared to a national average of $4,270. Over the war years the average debt of a cornbelt farmer increased about 50 per cent, that of an Iowa farmer, 69 per cent. Speculation was encouraged by the free flow of inflationary credit into the farm regions: a 10 per cent down payment would purchase a farm. Insurance companies and mortgage investment corporations provided funds for 40 per cent of the loans; the balance came from private investors or local banks. These latter institutions, almost as "wildcat" as in the days of Jackson, often had inadequate capital and floated loans too liberally. Iowa in 1920 had 1,763 banks, more than 20 per county, and the state could make the dubious boast of having more active banks than any other state.[24]

The price break of 1920 sent tremors through the shaky structure of farm finance, but did not initiate a serious collapse. Most farmers were able to gather enough resources to struggle through the decade. New loans could be floated for renewal of debts and most after 1920 were for that purpose. Since renewal loans were usually contracted with investment corporations, the percentage of debt held by creditors outside the cornbelt increased. Many of the county banks, weak institutions in the first place, failed to weather the economic changes. Forty-eight per cent of the nation's bank failures in the twenties were in Minnesota, Iowa, Missouri,

[24] Jones and Durand, 80-82; Case, *Agricultural History*, 174-75.

Nebraska, Kansas, and the two Dakotas. In Nebraska, for example, the number of banks decreased 51 per cent between 1920 and 1932. Caused in part by this failure of local credit, the number of foreclosures crept steadily upward. In thirteen Iowa townships where foreclosures had been negligible between 1915 and 1920, about 14 per cent of all loans canceled were by forced sale in 1925.[25]

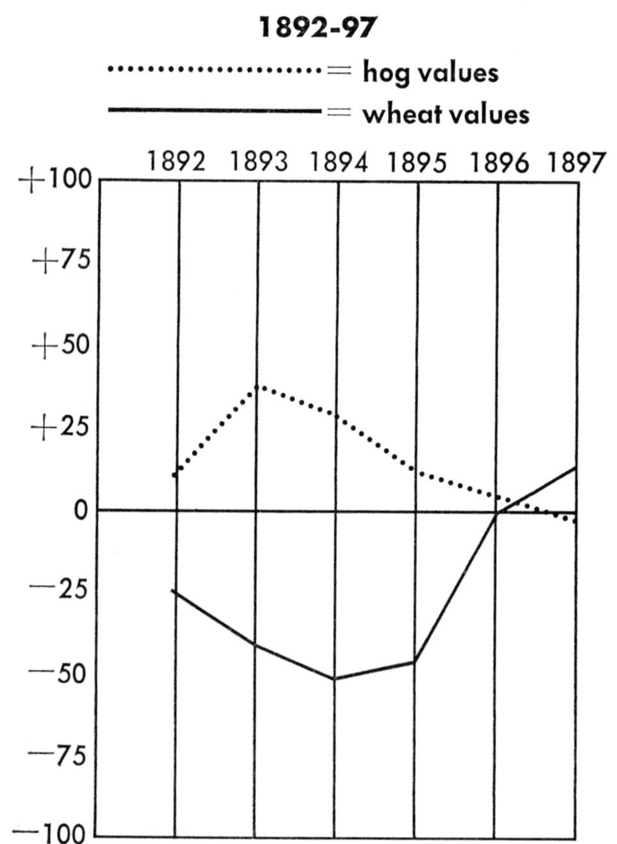

[25] Case, *Agricultural History, loc. cit.;* Soth, 111, 118; Filley, "Effects of Inflation and Deflation upon Nebraska Agriculture," 100.

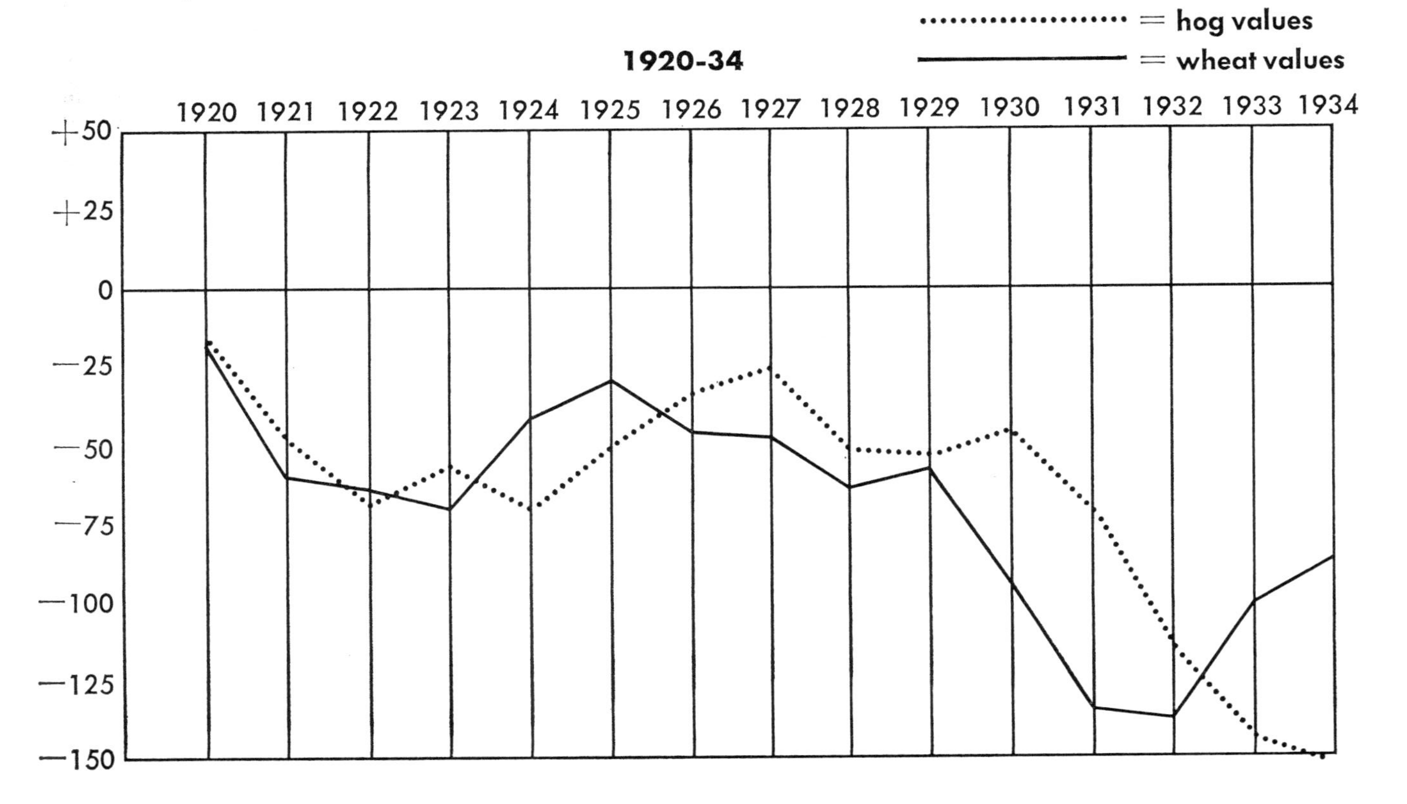

Comparison of Percentage Deviation of Year's Commodity Price from That of Preceding Year,
1920-34
= hog values
= wheat values
1920
1921
1922
1923
1924
1925
1926
1927
1928
1929
1930
1931
1932
1933
1934
+50
+25
0
—25
—50
—75
—100
—125
—150

The problem of farm credit became a crisis with the crash of 1929. The farmer's debt position was vulnerable and when investors and bank account creditors were forced to make calls upon their assets, investment institutions in turn had to press demands upon their farmer debtors. In eleven Iowa counties surveyed, more than half of the foreclosure judgments were in favor of insurance companies. Beyond this, the weak county banks were even more beleaguered: a total of fifty-seven failed in Iowa in 1930 and 161 more closed their doors in 1931.[26]

Between 1921 and 1933, 13 per cent of Iowa farm land was sold at foreclosure. Yet at the end of 1932, one billion dollars of debt was still outstanding on 45 per cent of the land in the state. As the accompanying chart indicates, foreclosures and bankruptcies were most frequent in the west north central farm states; the Iowa total of seventy-eight for every 1,000 farms in 1933 was the highest in the nation. Foreclosures averaged sixty-five for each

Number of Farms Changing Ownership by Foreclosure of Mortgages and Bankruptcy per 1,000 of All Farms, by Area and State[a]

	1931	1932	1933	1934
UNITED STATES	18.7	28.4	38.8	28.0
New England	6.3	10.3	13.2	12.8
East North Central	19.3	27.8	38.3	27.8
South Atlantic	19.4	16.1	32.2	22.5
East South Central	15.9	24.6	36.4	24.7
West South Central	16.3	27.0	35.2	22.1
Mountain	22.6	27.0	33.6	28.7
Pacific	19.6	26.8	36.0	30.6
WEST NORTH CENTRAL	25.8	43.8	61.5	44.4
Minnesota	31.2	42.9	59.1	37.5
Iowa	24.8	52.5	78.3	54.3
Missouri	23.7	42.1	51.2	36.1
North Dakota	34.1	54.0	63.3	31.3
South Dakota	33.2	49.2	78.0	64.2
Nebraska	21.8	34.4	58.2	45.8
Kansas	20.0	36.0	52.7	48.0

[a] U.S.D.A., *The Farm Real Estate Situation, 1933-34*, Circular 354 (Washington, 1935), 31.

county. Land values had declined so greatly (from $140 per acre in the late twenties to $92 per acre in 1932) that proceeds from a forced sale usually did not cover the full amount of the

[26] Case, *Agricultural History, loc. cit.;* Murray and Bentley, in *Black et al,* 162-65; *Iowa Union Farmer,* January 13, 1932.

debt. As a result, the debtor was left with a deficiency judgment to cover the balance. In 1921, 26.5 per cent of foreclosures in eighteen sampled Iowa counties ended with a bid for less than the amount of the debt; in 1932, 74 per cent carried a deficiency judgment.[27]

Even when the crisis was at its worst, about half of the farmers were mortgaged, another third carried debts not burdensome, and only one-sixth were hard-pressed. Farmers had been unduly optimistic in debt expansion, but they were not to blame for the failure of the international market or the assist given this failure by the Smoot-Hawley tariff. Neither were they responsible for the economic crash of the early thirties. The farmer debtor was a speculator, but not necessarily a greedy one whose only objective was to make a killing through increments in land value. Many were young farmers, thrifty enough to have saved money a few years earlier to buy a farm of their own.[28] A farmer even remotely facing foreclosure action might well contemplate that he was ensnared in an economic system that had tempted him with false promises, offered him easy credit, and then abandoned him in time of need. Farmer direct action was most vigorous in the attempt to halt forced sales; these actions occurred when foreclosures were the highest in number during the entire century and in the area where they were most frequent.

The American farmer has never been politically inarticulate. He did not remain a silent spectator to the creeping economic paralysis of the twenties. During the war period, membership in voluntary associations increased rapidly. Despite the disappearance of wartime prosperity, total family membership in the principal general farmers' organizations continued to increase from 1920 to 1930. The American Farm Bureau Federation, the largest of these, was founded only in 1919 and by 1920 claimed 317,000

[27] Murray and Bentley, in Black *et al*, 159; Howard W. Lawrence, "The Farmers' Holiday Association in Iowa, 1932-33" (unpublished M.A. thesis, State University of Iowa, 1952), 5-6; Soth, 122.

[28] Case, *Agricultural History*, 174. The average age in 1933 of 34 former members of the Farmers' Holiday Association responding to my questionnaire was 43.2 years. The average age of 84 arrested for Holiday activity at Denison, Iowa in April, 1933 was 43.4. This would suggest that many of the protesting farmers were among the age group that began farming during and shortly after the war.

members; despite sharp drop offs in the Midwest, membership by 1930 had increased slightly to 321,000. The Farmers' Union, founded in 1902, expanded into the corn and wheat belts during the war. Membership losses in the twenties were greater than for the Farm Bureau or Grange since the Farmers' Union represented a lower economic strata of farmers; even so in 1933 membership in the midwestern states was 60 per cent above the 1914 figure. Grange membership actually increased by 27 per cent between 1920 and 1930 as the organization reactivated its political interest and campaigned for a federal farm program. While probably no more than a fourth of all American farmers belonged to one of these associations, the organizations were the major political voice of agriculture and exerted considerable political pressure through the strong lobbies each maintained in Washington. A bipartisan farm bloc in Congress that included such influential legislators as Senators George Norris of Nebraska, Arthur Capper of Kansas, and Burton Wheeler of Montana made farm legislation one of the crucial political issues of the twenties.[29]

A host of legislative proposals attempted to check the ebbs and flows of prices and assure farmers a "fair exchange value" for their products. All of the alternatives debated in the thirties originated in the prior decade: the guaranteed cost of production scheme of the Farmers' Union, the complex export debenture plan of the National Grange (which would subsidize farmers for export losses through payment of negotiable "debenture" certificates), and domestic allotment, a planned relationship between production and demand. The farm bloc was sufficiently strong to garner for the farmer a few legislative fringe benefits such as the Capper-Volstead Act of 1922 which freed farmer cooperatives from the restrictions of the anti-trust laws and the Credit Act of 1923 which guaranteed farmer credit on short-term loans.[30]

The McNary-Haugen bill, the most important farm legislation proposed in the twenties, was widely accepted because it made use of traditional devices including a high tariff to protect American producers from foreign competition. A "two price system,"

---

[29] Robert L. Tontz, "Membership of General Farmers' Organizations, United States, 1874-1960," *Agricultural History*, XXXVIII (July, 1964), 147; Angus Campbell *et al*, *The American Voter* (New York, 1960), 414.

[30] Ross M. Robertson, *History of the American Economy* (New York, 1955), 381.

one for the domestic market, another for the world market, was reminiscent of the Grange's export debenture plan. The attempt to guarantee farmers fair exchange value suggested to the Farmers' Union leaders their cost of production proposal. A fair exchange value would be determined for each farm commodity based upon the ratio of current farm prices to purchasing power in 1910-14. The "fair" price was to be maintained, first through the protective tariff and second, by a government-chartered corporation that would buy a sufficient amount of each commodity to force its price up to the fair value level. Surpluses acquired by the corporation would be "dumped" abroad at the prevailing world price. Losses would be recouped through an equalization fee assessed on those farmers who sold for the higher prices on the domestic market. The farmer would gain to the extent additional income resulting from higher prices exceeded the equalization fee.[31] The McNary-Haugen bill was supported by all three major farm organizations and it passed Congress both in 1927 and 1928. Twice it was vetoed by President Coolidge. The Smoot-Hawley tariff and the resulting paralysis in international trade made dumping abroad unfeasible and rendered obsolete both the export debenture and the McNary-Haugen plans.

The Agricultural Marketing Act of 1929, the Republican answer to the farm problem, attempted to ameliorate the crisis by means of giant cooperative institutions. The act created a Federal Farm Board of eight members that was authorized to support new and existing farm cooperative marketing associations, obtain warehouses, and make advances to cooperating farmers. To finance the operation, Congress appropriated $500 million for a self-sustaining revolving fund. "Stabilization corporations," owned by the cooperatives, would use the fund to buy up surplus and carry on price support operations. With the board's encouragement, major grain cooperative associations, including the large Farmers' Union Terminal Association of Minneapolis, consolidated to form the Farmer's National Grain Corporation. For the first time in

[31] Theodore Saloutos and John D. Hicks, *Agricultural Discontent in the Middle West, 1900-1939* (Madison, 1951), 377-78; Robertson, 381. For a text of the McNary-Haugen bill see *Documents of American History*, ed. Henry Steele Commager (7th ed.; New York, 1963), II, 210. For a lively account of the fight for the bill that particularly emphasizes the opposition of Secretary of Commerce Herbert Hoover, see Arthur M. Schlesinger, Jr., *The Crisis of the Old Order* (Boston, 1957), 107-9.

history the federal government was committed, although indirectly, to stabilizing farm prices.

Had the Federal Farm Board been able to operate in a more placid economic climate it might have borne fruitful results. Mounting surpluses after 1929 soon glutted the stabilization corporations and the board itself was forced to engage in buying operations, only to suffer heavy losses when prices dropped. The board still held a gigantic backlog of unsold farm surplus when it was liquidated in 1933. There were other disabilities. Members of cooperatives were penalized by the operation of the act for they alone paid fees while any general price benefits would accrue to all farmers, not merely to those who paid the costs. Moreover, the Farm Board could not help all farmers. Cooperative techniques were most feasible when applied to non-perishables such as grain or wool but could offer little assistance to corn or livestock producers. The most severe decline in corn and hog prices did not occur until autumn of 1931 and by that time, Farm Board funds were already heavily committed in wheat, cotton, and wool support programs. When disaster struck the cornbelt farmer in 1931, the only federal agency that could have rendered him assistance was devoid of funds and ill-adapted to his needs.[32]

The legislative battles and defeats of the twenties were a political education for the revived farmer organizations. One indication of quickening political interest was the formation of the Corn Belt Committee in May, 1925. This was a loose confederation of twenty-four separate farm groups established to achieve harmony on a federal agricultural program and bring pressure for its approval. Initiative for its formation came from the National Farmers' Union where the idea was first suggested by Milo Reno, president of the Iowa Union. Chairman was William Hirth, president of the Missouri Farmers' Association, and during its seven-year life span the committee included among its members Henry A. Wallace, editor of *Wallace's Farmer*, and leaders of the Farm Bureau Federation in Nebraska, Indiana, and Iowa.[33]

---

[32] Robertson, 382; Saloutos and Hicks, 404-28; Murray R. Benedict and Oscar C. Stine, *The Agricultural Commodity Programs* (New York, 1956), 187.

[33] Benedict, *Farm Policies*, 220-21; Milo Reno, MS of address at Yankton, South Dakota, November 17, 1933; *Farmers' Union Herald* (St. Paul), May 18, 1931.

The committee endorsed, in general, the principle put forward by the Farmers' Union that farmers were entitled to a minimum of production costs as an equitable return for their products. More specifically, the organizations in the Corn Belt Committee joined forces to campaign for the McNary-Haugen bill and acted as an active lobbying force in the second unsuccessful attempt to make it law. The second veto and the passage of the Agricultural Marketing Act in an atmosphere of deepening crisis undermined the harmony of the committee. One group, led by Farm Bureau representatives, wanted to cooperate with the Farm Board and continue to press through orderly legislative processes for improved legislation. Another, headed by Hirth and Reno, consisting largely of representatives from areas that stood to profit little from the Republican farm legislation, opposed the board and urged a vigorous political drive for immediate cost of production prices. As a last resort this faction would have farmers withhold produce from market as a pressure tactic. With the committee dominated by the radicals, all of the moderates seceded en masse in 1931, ending the career of the Corn Belt Committee.[34]

The Farmers' Union inspired the Corn Belt Committee in the twenties and initiated most of the radical agrarian legislation of the thirties. With the collapse of the Corn Belt Committee, this organization became the battleground for conflicting ideas on what political course farm organizations should pursue.

The Farmers' Educational and Cooperative Union was founded by "Newt" Gresham of Texas in 1902. During the war years the organization expanded outside the cotton regions and by 1928 its principal strength was in Nebraska, the Dakotas, Kansas, Oklahoma, Iowa, Montana, and Colorado. In its original program the Union was committed to such broad policies as uniform prices for all commodities, encouragement of scientific farming, and opposition to the mortgage and credit system. In the twenties the Union extended its interests into the cooperative field. Largest of union cooperatives was the Farmers' Union Terminal Association of Minneapolis, servicing members in the wheat states of Minnesota, North and South Dakota. On a smaller scale, state associations like the one in Iowa operated an insurance company

[34] *Farmers' Union Herald*, May 18, 1931.

and livestock commission houses at Sioux City, South St. Paul, and Chicago.[35]

The original objectives of the Farmers' Union contained no provision relating to political action, but from the very beginning a minority faction urged upon the group an active legislative program and some called for guaranteed cost of production prices for farmers. In the early twenties Milo Reno took up the cost of production cry and pleaded with state and national officers to call together all organizations committed to that principle to design a legislative program. This proposal led to the formation of the Corn Belt Committee.[36]

Cost of production could be a slogan like "free silver at 16-1" which might mean all things to all people or it could represent a fairly specific legislative program. In the latter form, it meant an itemizing of the average expenses of producers of each commodity, including an allowance for wages to the farm operator, and arriving at a price for products consumed domestically that would return to this average operator his costs, labor, and a reasonable profit. Processors would be required to purchase at this price the percentage of the year's crop that would be consumed on the home market. Any surplus was the farmer's own responsibility: to be stored in prospect of better times or disposed of at world market prices.[37]

The Farmers' Union was far from unanimous in subscribing to cost of production. State associations enjoyed considerable autonomy and national membership fluctuated on a yearly basis as various state units seceded and rejoined over issues revolving around finances, policies, and personalities. For example, the large Nebraska Union was dedicated to a conservative policy of using the cooperative principle to return to a self-regulating farm economy, eliminating bureaucracy and lowering tariffs to a competitive level. Its leaders opposed the cost of production political

---

[35] W. P. Tucker, "Populism Up to Date: The Story of the Farmers' Union," *Agricultural History*, XXI (October, 1947), 203; Wesley McCune, *The Farm Bloc* (Garden City, N.Y., 1943), 193-95; *Farmers' Union Herald*, May 18, 1931. The best available data gives Farmers' Union membership as 153,624 family members in 1917; 131,475 in 1919; 77,953 in 1933. Tontz, 155.

[36] Reno, MS of speech at Yankton, November 17, 1933 and MS, "For Miss Prescott" (n.d.).

[37] *Iowa Union Farmer*, July 28, 1927.

program. The *Farmers' Union Herald,* voice of the Union grain interests of the Northwest, remained lukewarm. The Agricultural Marketing Act drew a clear line of demarcation between political activists and cooperators and prompted an internal feud that changed the course of the organization.

Dissent was already present when Charles Barrett, a founder and long time president, retired in 1928 and was replaced by C. E. Huff, president of the Kansas Union. Huff was an ally of M. W. Thatcher, who headed the Farmers' Union Terminal Association, largest of the business enterprises. The Federal Farm Board promised real benefits to the grain cooperative, hence Huff and Thatcher were eager supporters. An opposition group, headed by Milo Reno and E. E. Kennedy of the Illinois Union, urged combining with cooperative business activities a determined political drive for cost of production legislation. Reno did not repudiate completely the cooperative idea; his Iowa Union engaged in sizeable business activities. Nevertheless, his organization, acting independently of the national, continued to press for cost of production. This conflict in purpose incited open controversy. Managers of the Union Livestock Commission House at Sioux City, allied with the national officers and the Nebraska Union, charged that Reno squandered cooperative funds for political purposes and in an acrimonious battle succeeded in removing the business from the control of the state Union leaders.[38] Discords burst into the open at the convention of November, 1930 and the political activists were able to muster enough support to defeat Huff and elect to the presidency John Simpson, president of the Oklahoma Union. Simpson had been outside the arena during most of the political infighting since the Oklahoma Union had only recently been reconciled to the national after several years of estrangement. Nevertheless, Simpson was elected as an anti–Farm Board candidate; through his victory the Farmers' Union committed itself to a program of political action.

The triumph of the radicals fanned dissension further. In Minnesota the Union split into two rival groups and in Iowa a small "bolter" faction joined with the cooperators in a policy at odds with Reno's state organization. To protect a valuable business from ravishing by what they considered a reckless political

---

[38] *Ibid.,* January 29, 1930.

element, Thatcher's Farmers' Union Terminal Association was absorbed into a larger enterprise, the Farmers' National Grain Association, and Huff became president.[39] E. E. Kennedy charged that the Farm Board revolving fund was being used to coerce organizations into the Hoover farm program and that at the recent convention liberal loans had been promised to Unions in states that would support Huff. For their part, the business faction charged that political leaders, particularly Reno, dipped into cooperative funds for personal and organizational expenses—a charge not entirely false.[40] The bitterness of this fight arose from implications by each side that their opponents were guilty of financial mismanagement or even worse, bribery and corruption. The eventual emergence of the Farmers' Holiday Association as the offensive arm of the political action group was in part an outgrowth of this struggle. An independent organization, the Holiday Association would not embarrass the extensive business operations in which the parent Farmers' Union was engaged.

John Simpson and Milo Reno, the two principal leaders who emerged at the head of the Farmers' Union political offensive, were both seasoned veterans of farm struggles and farm organization politics. Both had behind them ten years of experience in responsible and successful leadership at a state level. Each had almost single-handedly built the Farmers' Union organization in his state.

John Simpson, born in 1871, was a native of Oklahoma City and an attorney by profession. In appearance he was a grandfatherly man with benign features, but he was a spell-binding orator as a Farmers' Union leader had to be. During the war he was criticized for dissent, but he emerged as one of the most successful Farmers' Union organizers, building the Oklahoma Union from 231 to 23,000 members. He was a Populist, long an advocate of monetary inflation. He supported the McNary-Haugen bill, joined with Reno in backing cost of production and opposed Herbert Hoover in 1928. As president of the Farmers'

[39] *Ibid.*, July 29, August 23, 1931. The Farmers' National Grain Corporation was one of the most important creations of the Federal Farm Board. It combined into one various grain-marketing cooperatives; the largest stockholder was the Farmers' Union Terminal Association. Tucker, 206-7.

[40] John Chalmers (former president, Iowa Farmers' Holiday Association), personal interview, October 21, 1961.

Union he devoted a major portion of his attention to the political program and he made frequent appearances before the Senate and House committees on agriculture. He met with President-Elect Roosevelt at Hyde Park in 1932 to discuss farm legislation.[41]

Milo Reno was almost a caricature of the evangelical type of farm leader. He was an orator who embellished his speeches with homely farm analogies and liberal invocations of Biblical writ. A flaming red necktie was his trademark when he appeared at country gatherings and he wore expensive ten gallon hats. He chain smoked cigarettes, fiddled at country barn dances, and had a reputation as a "ladies man" that still causes a chuckle among old time Farmers' Union members.

Reno was born in Wapello County, Iowa in 1866, received his education in a log school house and studied for the ministry a short time at Oskaloosa (Iowa) College. He never held a charge, but he was an ordained Campbellite minister and often preached from rural pulpits. He was associated with most of the rural protest movements that had risen during his sixty-six years. His family had supported the Greenback party and as a boy Reno studied Greenback tracts in his home at Agency, Iowa. His family voted for Weaver in 1880 and for Ben Butler in 1884. In 1888 he campaigned for the Union Labor party, voted the Populist ticket in 1892 and took an active part in the Bryan campaign.[42]

Milo Reno joined the Farmers' Union in 1918—a turning point in his career. Within two years he was secretary-treasurer of the state organization and in 1921 was elected president. Until his death in 1936, Reno dominated the Iowa Farmers' Union. He resigned the presidency in 1930 to assume full-time direction of the Union's insurance business, a job that commanded a $9,600 annual salary, but he remained nonetheless the *de facto* head of the organization. Under his leadership the Iowa association grew

---

[41] Gilbert C. Fite, "John Simpson: The Southwest's Militant Farm Leader," *Mississippi Valley Historical Review*, XXXV (March, 1949), 563-84; ed. Gilbert C. Fite, "Some John A. Simpson–Franklin D. Roosevelt Letters on the Agricultural Situation," *Chronicles of Oklahoma*, XXVI (Autumn, 1948), 336-45.

[42] Reno, "For Miss Prescott"; Roland A. White, *Milo Reno: Farmers' Union Pioneer* (Iowa City, 1941), 22-23; Dale Kramer, *The Wild Jackasses: The American Farmer in Revolt* (New York, 1956), 191-98. Kramer's popular account is an unusually good source since the author was a personal friend of Milo Reno and served for three years as editor of *Farm Holiday News*.

to include about 10,000 members in 1929. In 1920 Reno won the endorsement of the Iowa Union for a principle that remained its stock in trade for fifteen years: cost of production prices for farmers. "If the Farmers' Union means anything on earth," he declared on the floor of the convention, "it means the right of you and me to determine the value of the products of our labor, just as organized labor, organized manufacturing, or organized banking."[43]

As has been said of William Jennings Bryan, Milo Reno was "a boy who never left home."[44] "I have been a greenbacker for over half a century," he wrote to Congressman Ernest Lundeen in 1933. In the last speech he ever delivered to the National Farmers' Union he recalled sentimentally, "As a boy I lived in the adjoining county of the one in which James B. Weaver lived. In fact I sat at the foot of that great statesman, philosopher, Humanitarian and Christian and learned what government really meant."[45] Reno was foremost an inflationist, for only with an expanded currency could the economy pay cost of production prices. In a favorite analogy, he envisaged two little boats, one of gold, one of pine, bearing wheat across the imaginary river of trade. The wheat borne in the gold boat had no more value than wheat transported in the pine boat. Why should the material of which the boat was made effect in any way the value of the wheat? The crisis of the twenties was not caused by overexpansion or market contraction, rather it was the result of "the most hellish program ever conceived in the minds of men": the deflationary program of the Federal Reserve Board that in eighteen months destroyed thirty billion dollars of farm values.[46] This was typical of Milo Reno. In place of the impersonal workings of an economic system he was prone to find deliberate malfeasance of individual men. The attacks he levied against the Federal Farm Board and its chairman, Alexander Legge, parallel almost exactly

---

[43] White, 35.

[44] Richard Hofstadter, *The American Political Tradition* (Vintage ed.; New York, 1954), 194.

[45] Reno to Lundeen, October 8, 1933; National Farmers' Union, "Minutes of National Convention, Kankakee, Illinois, November 19, 1935," Milo Reno collection, State University of Iowa.

[46] Reno to Roy Hildebrand, August 28, 1933; Reno, interview in *Yale Daily News* (New Haven, Connecticut), January 6, 1934.

the charges he would make after 1933 against another farm program and other leaders.

On July 6, 1927, Reno introduced to the Corn Belt Committee the following resolution: "If we cannot obtain justice by legislation, the time will have arrived when no other course remains than organized refusal to deliver the products of the farm at less than production costs." [47] This proposal, the first suggestion of the idea that gave rise to the Farmers' Holiday Association five years later, attracted little attention while the McNary-Haugen bill was still a viable political issue and farmers had not yet suffered the disaster of 1929.

With the early thirties and the simultaneous approach to ruin of both farm markets and farm prices, a midwestern farmer could look back over the political efforts of a decade and find only the ghost of the defeated McNary-Haugen plan, made obsolete by prohibitive tariff barriers, and an impotent Federal Farm Board helpless to provide him needed assistance. A radical plan for withholding of farm products, drastic by 1927 standards, might now seem a possible, even a desirable measure.

---

[47] Reno, "For Miss Prescott."

Two

UNCLE REUBEN AT THE CROSSROADS

When the international harvester people need some money to buy more diamonds or poodle dogs for their wives they just add a dollar or two to the price of a harvester and Uncle Reuben at the crossroads pays the extra price [Robert Moore, organizer, Iowa Farmers' Union, quoted in *Social Forces,* XII (March, 1934), 370].

Farm prices began a downward spiral in April, 1931 that was unchecked until June, 1932 when across the board a farmer's income had dwindled to half that of the preceding year. Hogs sold for 3¢ a pound, cattle for 5¢, and corn for 10¢ a bushel. Within this interval the first direct action movement appeared in Cedar County, Iowa and by the summer of 1932 a plan for organized withholding of produce from market was taking shape in Iowa.

State veterinarians attempting to enforce Iowa's law for compulsory innoculation of dairy cattle against tuberculosis were assaulted in Cedar County in the spring and summer of 1931 and a thousand farmers from the area converged on the state capitol to demand repeal of the law. This county, where the President of the United States, Herbert Hoover, had been born in 1874, was in the midst of a prosperous farming region. The value of land per acre was the highest of any county in the state and as a corollary, average mortgage debt was the greatest in Iowa. Farmers were faced not only with the perennial problems of debt and falling prices, but the wave of bank failures that had spread across Iowa had taken a heavy toll in the county. The very day the first farmer protest meeting was held, the Cedar County Bank at the county seat, Tipton, closed its doors.[1] The compulsory tuberculin test was salt that stung the wound of economic discontent. A cow condemned as a reactor had to be committed to slaughter and even with partial compensation by the state and federal governments, the owner lost an average of $130 on each animal. The veterinarians were scapegoats for deeper frustrations, but the history of social movements is replete with examples of irrational protests that have distorted facts and attacked those who were neither real problems nor real enemies.[2]

Tubercular testing began in Iowa on a voluntary basis in 1919. A revised Bovine Tuberculosis Law passed in 1923 provided that a county would be eligible for enrollment under an Accredited Area plan which provided for compulsory testing of all cattle in the county once 75 per cent of cattle owners signed petitions. The law was extended in 1929 when testing was made compulsory in every county where 50 per cent of cattle owners approved. These advances were for the most part willingly accepted in Iowa, but a determined minority contrived to place roadblocks in the way of the program.[3]

Cedar County was recognized as hostile territory. When attempts had been made in 1926 to enroll the county under the Accredited Area plan, a group of farmers led by J. W. Lenker of Wilton Junction had filed for an injunction, arguing that the law was unconstitutional, the number of signatures on the petition inadequate, and local hearings improperly conducted. The injunction was eventually upheld by the Supreme Court and the legislature remedied the legal defects with the compulsory testing law of 1929. Shortly thereafter, the opponents in Cedar and Mitchell Counties (in northern Iowa) filed injunctions to prevent the Department of Agriculture from carrying out the law. These

[1] Soth, 114, 117; Frank Dileva, "Frantic Farmers Fight Law," *Annals of Iowa*, XXXII (October, 1953), 88.

[2] Walter Davenport, "Get Away from Those Cows," *Colliers*, LXXXIV (February 27, 1932), 11; Kramer, 209-10. For further discussion of irrational factors in social movements see Talcott Parsons, "Social Strains in America," *The New American Right*, ed. Daniel Bell (New York, 1955), 127-28.

[3] Dileva, "Frantic Farmers," *Annals of Iowa*, 89; [David E. Archie], "Times of Trouble: The Cow War," *The Iowan* (Shenandoah, Iowa), VII (April-May, 1959), 29.

gestures delayed for more than two years any attempt to test cattle in Cedar County.[4]

Grass-roots opposition to testing was unscientific and perverse, but granting this, the farmers' complaints were not entirely baseless. No credence can be attached to assertions that testing caused cows to abort or that tuberculosis could not be transmitted from cows to humans. This was a kind of cant being broadcast in the area by Norman Baker, owner of Station KTNT, Muscatine, a charlatan and imposter whose guaranteed "cure for cancer" eventually merited him a federal prison sentence.[5] Opposition to testing was greatest in areas reached by Baker's voice. Nevertheless, the tubercular test was not 100 per cent accurate and opponents widely publicized the minority of instances where cows testing positive were discovered upon slaughter to be pure.[6] A court ruling in Mitchell County in 1929 referred to the fact that 9 to 14 per cent of reactor cows proved on slaughter to be uninfected and held seizure of reactors to be a violation of property rights.[7] Farmers had difficulty understanding why animals whose milk was impure according to the test could nonetheless be sold to packers and the meat marketed. Had the Iowa Department of Agriculture preceded the compulsory testing with a vigorous educational program the troubles might have been avoided. Mark Thornburg, at the time Iowa Secretary of Agriculture, looking back in retrospect thirty years, believed attempts to test in the hostile areas had been premature. As it was, in the depression year of 1931, when state veterinarians, strangers in Cedar County, appeared on farms to conduct a test in which farmers had no confidence and which could result in a loss of several hundred dollars a farmer could ill afford, the interlopers were greeted with sticks, stones, and mob resistance.[8]

---

[4] Archie, *The Iowan,* 29-30.

[5] Warren B. Smith, "Norman Baker—King of the Quacks," *The Iowan,* VII (December-January, 1958-59), 16-18.

[6] Actually, the cattle in Cedar County when subjected to test during the three years from 1923 to 1926 had proven purer than the state average. In the county, out of 26,000 animals examined, 2.80% were reactors compared to the state average of 5.72%. Archie, *The Iowan,* 34.

[7] Loftus *et al* v. Thornburg, District Court for Mitchell County (Iowa), 1929. This ruling is reproduced in full in *Iowa Union Farmer,* October 23, 1929. It was not sustained by the Supreme Court of Iowa.

[8] Archie, *The Iowan,* 33; George Ormsby (Wilton Junction, Iowa), personal interview, August 23, 1961.

The Department of Agriculture chose to bring the matter to a head in February, 1931. Notices to have dairy herds tested within fifteen days, the first served in Cedar County, were sent to the individuals who had been the leaders of the protest. Veterinarians tested the herd of William Butterbrodt on the morning of March 5, but when they arrived at the farm of E. C. Mitchell, south of Tipton, a mob of 200 farmers prevented the test. When they returned four days later to read the results of the test at the Butterbrodt farm, 500 "neighbors" barred their progress.[9] The same week a farmer meeting in Tipton sent to the governor a petition demanding repeal of compulsory testing and organized a Farmers' Protective Association. President of the Association was Jake Lenker, who had organized the legal opposition a few years earlier, and one of the prime movers was Paul Moore, an organizer for the Farmers' Union and brother to the state secretary of that organization. On March 19, 1,000 southwest Iowa farmers, organized by the Protective Association, descended on the state capitol in Des Moines. They packed the legislative galleries as the Animal Husbandry Committee held public hearings on a voluntary testing bill, introduced by a friendly representative. No opponents of the bill braved an appearance. The farmers had their day in court and that was about all the Republican state administration was willing to grant them. Governor Turner expressed no sympathy with voluntary testing and asserted it was his responsibility as state executive to enforce the law. The voluntary testing bill was defeated a few days later.[10]

The Iowa Farmers' Union encouraged, but did not actively sponsor, the protest against compulsory testing. Union organizers and members joined the Protective Association, but the two groups remained separate probably because the Union could not risk having its business enterprises jeopardized by potential law suits.[11] Nevertheless, the *Iowa Union Farmer* publicized the movement and some of the responsibility for stirring opposition to testing in Iowa rests with Milo Reno. Public health measures were fine, he asserted in 1927, but he opposed "political preferments" being granted to "pap-stickers." This was a colorful way

---

[9] *Iowa Union Farmer,* March 11, April 8, 1931; Archie, *The Iowan,* 30.

[10] *Iowa Union Farmer,* March 25, 1931; Dileva, "Frantic Farmers," *Annals of Iowa,* 81-96; Archie, *The Iowan,* 31.

[11] White, 51-52.

of expressing the real issue. The Union resented the test's sponsorship by its rival organization, the Farm Bureau Federation, and its administration by county agents and the state agricultural college at Ames, both politically and financially linked to the Farm Bureau. The real issue in Cedar County, Reno declared, "lies in the fact that their property is no longer their own. Any little shyster who has come out of a certain college in this state can go to a farmer's property and conduct a test which is more apt to be wrong than right." [12]

The State Department of Agriculture was determined to make a test case in Cedar County. On August 5 injunctions were obtained restraining Jake Lenker and forty-four other leaders of the Protective Association from interfering with further testing. Two weeks later, Dr. Malcolm, the state veterinarian, attempted a test near West Liberty. He and his assistants were pelted with water and eggs by an armed mob. In nine attempts that day only one test was completed. Three farmers were arrested and Governor Turner authorized the swearing of sixty-five deputies. On September 21, the state agents passed the calves and waved the red flag directly before the bull: they attempted to test at the farm of Jake Lenker, just outside the village of Wilton Junction. In the battle that resulted, 450 farmers, armed with clubs and throwing mud and Irish confetti, fought through clouds of tear gas to compel the deputies to retreat. Several were injured, the state officials' automobiles were smashed, and Dr. Malcolm himself was painfully bruised.[13]

Governor Turner quickly dispatched three regiments of the Iowa National Guard, a total of 2,000 men, into the county. For the first time in the history of the state, machine guns were set up along dirt country roads and soldiers patrolled the rolling pasture land of Cedar County. Several leaders of the Protective Association were arrested and testing proceeded backed up by military power. Within a week over 14,000 head of cattle, including Lenker's herd, had been tested and troops moved to neighboring Henry, Muscatine, Jefferson, and Des Moines Counties to complete the job. Lenker and Paul Moore were sentenced to three-year prison terms on charges of conspiracy to violate the

[12] *Iowa Union Farmer,* February 16, 1927. Reno is quoted in Saloutos and Hicks, 439.

[13] Dileva, "Frantic Farmers," *Annals of Iowa,* 100-102.

Iowa tubercular law; both eventually served about one month. Martial law was never declared in southeast Iowa, but the National Guard remained for two months, until testing had been completed.[14]

The Cow War was an isolated event, but it was a harbinger of the rebellious spirit depression had set astir in the countryside. If the trouble in Cedar County gave any indication of what form future unrest might take, it was a revolt of traditionalism against modernism and change. It was initiated by farmers more prosperous than most of their fellows, but beset by critical economic conditions. It was irrational in form and chose scapegoats as enemies; it displayed little internal discipline and was easily roused to violence. It collapsed quickly when force was brought into play. The Farmers' Protective Association, like the later Holiday Association, was an independent adjunct of the Farmers' Union.

While Cedar County farmers spent themselves in futile protests against interloping veterinarians, the Iowa Farmers' Union was working to direct the discontent of which the Cow War was a symptom into more organized channels. The Iowa Farmers' Union claimed 9,600 members in 1932; it was outnumbered by the Farm Bureau in its home state and it espoused a program to which the national organization did not entirely subscribe. Face to face contacts were so frequent and personal relations so intimate within the Iowa Union that it had more the characteristics of a lodge or fraternal organization than of a political group. Since in addition to county organizations there were from three to five locals in each county, many of the members held leadership positions. Decisions at the state level were by democratic vote and the annual convention often attracted several thousand.[15] Its newspaper, the *Iowa Union Farmer,* was a vigorous, homespun journal rich in anecdotes and personal news; the editor was H. R. Gross.[16] Milo Reno was on a first name basis with probably

[14] *Ibid.,* 103-7; Archie, *The Iowan,* 34-35, 52-53.

[15] *Iowa Union Farmer,* September 23, 1931. In 1935 when membership was depleted by years of financial hardship, the Iowa Farm Bureau still had 24,202 members. At the height of its strength in 1929, the Iowa Farmers' Union had 10,063. *Iowa Union Farmer,* March 27, 1929, April 6, 1935.

[16] Gross is currently a Republican member of the House of Representatives from the 3rd Iowa district.

a majority of the members and the members in their turn felt free to appeal to Reno for personal counsel or to write frequent letters outlining political and social views. Reno was religious in his correspondence duties and his letter files for 1933 indicate that he answered every one of the nearly 3,000 letters addressed to him. Opponents of Reno had been eliminated or isolated in the battle over cooperatives so at the heart of the Union's membership were Reno's steadfast friends.

The Iowa Farmers' Union was the organized core that led the cornbelt rebellion. Rural protest organizations have traditionally consisted of two elements: first, the formal organization with its officers, program, and institutionalized structure; and second, the movement element—a reservoir of disaffected and discontented people from whom the formal organization draws sustenance and support.[17] Within a year the small Iowa Farmers' Union inaugurated an insurgent movement that could sway state governments and influence federal legislation. The Union did not create sentiment, it crystallized a sentiment that was already present.

Milo Reno's resolution calling for a general withholding movement had slumbered for four years, with occasional references in the *Iowa Union Farmer* and a few endorsements by locals in the state. Support re-awakened as economic hardship pressed heavier in the cornbelt. John Bosch, president of the strong Farmers' Union in Kandiyohi County, Minnesota, urged the calling of a farm strike and at Reno's invitation addressed the convention of the Iowa Union in 1931. At the National Farmers' Union convention of that year, Reno and Bosch combined to present a resolution calling for a farm strike to begin January 1. The motion was beaten by a decisive margin of 68-19 delegate votes. Even though John Simpson was president and the political action group had control, the Union was engaged in substantial business operations and within the ranks there was some keen opposition to the entire political program. The Union did not spurn direct action, it simply passed the initiative from the parent body to Milo Reno and his core of followers from Iowa and Minnesota.[18] Although plans for

[17] Ray E. Wakeley, "How to Study the Effects of Direct Action Movements on Farm Organizations," *Social Forces,* XII (March, 1934), 380-82; Lipset, 60.

[18] *Iowa Union Farmer,* February 29, 1928, February 1, 1931. John Bosch, personal interview, April 1, 1962.

a withholding movement proceeded independently, there was always a close bond with the Farmers' Union. In every particular the program of Reno and his followers paralleled that of the Union. John Simpson and E. E. Kennedy, shortly to become national secretary, advised and sustained Reno in his independent resistance. In a real sense, the Farmers' Holiday Association was a strong-arm auxiliary of the Farmers' Union.

At Simpson's behest, forty Farmers' Union leaders, seventeen of them from Iowa, journeyed to Washington, D.C. in February, 1932. Ostensibly the purpose was to demonstrate support for two bills being considered by the Senate Committee on Agriculture and Forestry, but the real aim was to publicize the mounting discontent in the farmbelt and provide a sounding board for the Farmers' Union legislative program. The two bills, introduced by sympathetic senators, were distinctly Farmers' Union measures. The first, sponsored by Senator Lynn Frazier, former Non-Partisan League governor of North Dakota, called for a refinancing of farm mortgages by the federal government at a 1½ per cent yearly payment on principal and a 1½ per cent yearly interest. The second, introduced by Senator Elmer Thomas, a close friend of his fellow Oklahoman John Simpson, was titled "To Abolish the Federal Farm Board and Secure to the Farmer Cost of Production." Among those present from Iowa were Milo Reno, John Chalmers of Boone, Clinton Savery of Logan, and Paul Moore from Cedar County—men who would emerge within a year's time as leaders of midwestern farm insurgency. William Lemke of North Dakota, not yet a member of the House of Representatives but already a champion of inflationary credit, was principal witness for the Frazier bill. Implicit in the testimony of most of the witnesses was a threat that unless legislation was soon passed farmers were ready to resort to drastic measures. A representative of the Colorado Farmers' Union declared, "The farmers of this nation are still a conservative people. I emphasize the 'still.' All over this nation, however, conditions prevail which enable a busted farmer to gather the idea that if the government is against him and for every other kind of industry, he is possibly not going to be so conservative." Paul Moore, one of the principals in the Cow War, was even more emphatic:

Moore. I think the time will come when we have been driven far enough, when we will have what may be called a farmer's strike. . . .

I think that the next step that will be taken by the organized farmers of this country will be the strike.

Senator Frazier. You mean, unless legislation is passed.

Moore. Unless legislation is passed.[19]

Despite the friendliness of the handful of senators who made up the sympathetic subcommittee, a few breaths of the Washington atmosphere were sufficient to convince the visiting farmers that possibilities for legislative assistance in the Seventy-second Congress were hopeless. Milo Reno reported to his Iowa followers that the farm bloc was dead and that relief was in prospect for everyone except the farmers. "I am sadly afraid," he added, "it is going to be necessary to put many new faces in Congress before we can break the grip of Wall Street and international bankers on our government." John Chalmers bitterly relayed the words of Republican Senator L. J. Dickinson of Iowa who told him farmers were still driving cars and smoking cigars while railroads went broke. With a trite but biting metaphor, Chalmers urged, "Strike while the iron is hot!"[20]

Momentum was already gathering. The *Iowa Union Farmer* reported that almost to the man the delegation returned from Washington determined upon an immediate withholding movement. Glen Miller, president of the Iowa Union, coined a term when he declared that if banks could do it, why shouldn't farmers call a "holiday" where corn and meat and milk would be kept at home until senators and the public alike learned the importance of the men who tilled the soil. One rural bard, a better agitator than a poet, captured the spirit:

We can't continue longer now
Upon our weary way—
We're forced to halt upon life's trail
And call a "holiday."

Let's call a Farmers' Holiday
A Holiday let's hold
We'll eat our wheat and ham and eggs
And let them eat their gold.[21]

---

[19] U.S., Congress, Senate, Committee on Agriculture and Forestry, *Hearings on S. 3133, To Abolish the Federal Farm Board and Secure to the Farmer Cost of Production,* 72 Cong., 1 sess., 1932, 52 and *passim;* U.S., Congress, Senate, Subcommittee of the Committee on Agriculture and Forestry, *Hearings on S. 1197, To Establish an Efficient Agricultural Credit System,* 72 Cong., 1 sess., 1932, 60 and *passim.*

[20] *Iowa Union Farmer,* February 24, March 9, 1932.

[21] *Ibid.,* February 10, March 9, 1932.

Fifteen hundred Boone County, Iowa farmers gathered at the county fairgrounds on February 19 in response to a call from John Chalmers, president of the county Farmers' Union. They pledged to "stay at home—buy nothing—sell nothing!"[22] The Farmers' Holiday movement was underway.

Events preceding the preliminary organizing drive indicate that the withholding movement was primarily political in purpose, designed to apply pressure for cost of production and other legislation favorable to farmers. Its sponsors were the faction within the Farmers' Union that had been advocating a vigorous political program for three years. For Reno and Simpson this always remained the *raison d'etre*, but this purpose was not always clearly defined. Many local leaders and participants had a different idea of what should be achieved. John Chalmers, for example, was a veteran of the labor movement and had participated in strikes as a member of the carpenter's union. (His father had been blacklisted in the mine strike at Spring Valley, Illinois forty years earlier.) Chalmers believed that just as working men by their organized efforts had rationed the supply of labor in order to advance wages, so farmers by withholding produce could force up farm prices. Cost of production would be achieved by economic, not political means.[23]

From February until August the pages of the *Iowa Union Farmer* contained little else but plans for the forthcoming withholding action. Reno, Chalmers, Miller, and other Farmers' Union leaders toured the counties of western Iowa often addressing two rural gatherings each day. The organizers reached for support outside of the Iowa Farmers' Union. Reno cultivated Union associates in Wisconsin, South Dakota, and Minnesota; Miller consulted with officials of the Iowa Farm Bureau.[24]

The organizing drive culminated when 2,000 farmers assembled at the Iowa fairgrounds in Des Moines on May 3 to launch the Farmers' Holiday Association as a national farm movement. Iowans predominated but Minnesota, Missouri, Illinois, Montana, Wisconsin, Oklahoma, and Nebraska were represented. Milo

---

[22] John Chalmers, personal interview, October 21, 1961.

[23] Chalmers, personal interview. Most historical accounts have erroneously emphasized the economic rather than the political purposes of the farm strike and the Holiday Association.

[24] *Iowa Union Farmer*, March 9, 1932; White, 74.

Reno, the inevitable choice, was named president; John Bosch of Minnesota, the vice-president.[25] Implicit in the simple resolutions adopted was the assumption that the goals of the organization could be quickly and decisively achieved. A general withholding movement was to begin July 4 to continue for thirty days or until cost of production prices were realized. Left unanswered was the crucial question of whether this objective was to be achieved through political action or by economic pressure.[26]

Cost of production was a panacea. "Concede to the farmers production costs," Milo Reno prophesized, "and he will pay his grocer, the grocer will pay the wholesaler, the wholesaler will pay the manufacturer and the manufacturer will be able to meet his obligations at the bank. Restore the farmer's purchasing power and you have re-established an endless chain of prosperity and happiness in this country." [27] A reading of the many speeches of Milo Reno over a five-year period reveals not one instance in which he carefully explained and evaluated this principal tenet of the Farmers' Holiday Association.[28] Had he done so the result would have been approximately as follows: To compute cost of production figures tabulate all essential operating costs of farm business units and determine an average. Add an annual wage of $1,250 for the operator, a 5 per cent return on investment in land and a depreciation allowance. Then determine the percentage of the total farm product of the state derived from each farm commodity. For example, if hogs constitute 30 per cent of the farm produce of a state, then 30 per cent of the total production costs for each farm should be met by income from hogs. The market price for hogs which processors would be compelled by law to pay would be fixed accordingly.[29] To have realized returns equal to cost of production in 1932 a farmer would have had to receive

[25] *Iowa Union Farmer*, May 4, 1932.

[26] For a discussion of the characteristics of incipient protest movements see Wendell King, *Social Movements in the United States* (New York, 1956), 42-44.

[27] Reno, "Why the Farmers' Holiday?" Radio address of July 20, 1932, quoted in White, 151.

[28] For example, see the thirteen speeches reproduced in White, 121-92.

[29] E. E. Kennedy, Radio address of October 27, 1934, quoted in *Farm Holiday News*, November 1, 1934. The paper was published at St. Paul (February, 1933–January, 1934), Marissa, Illinois (February, 1934–December, 1935), and Ames, Iowa (December 26, 1935–August, 1936).

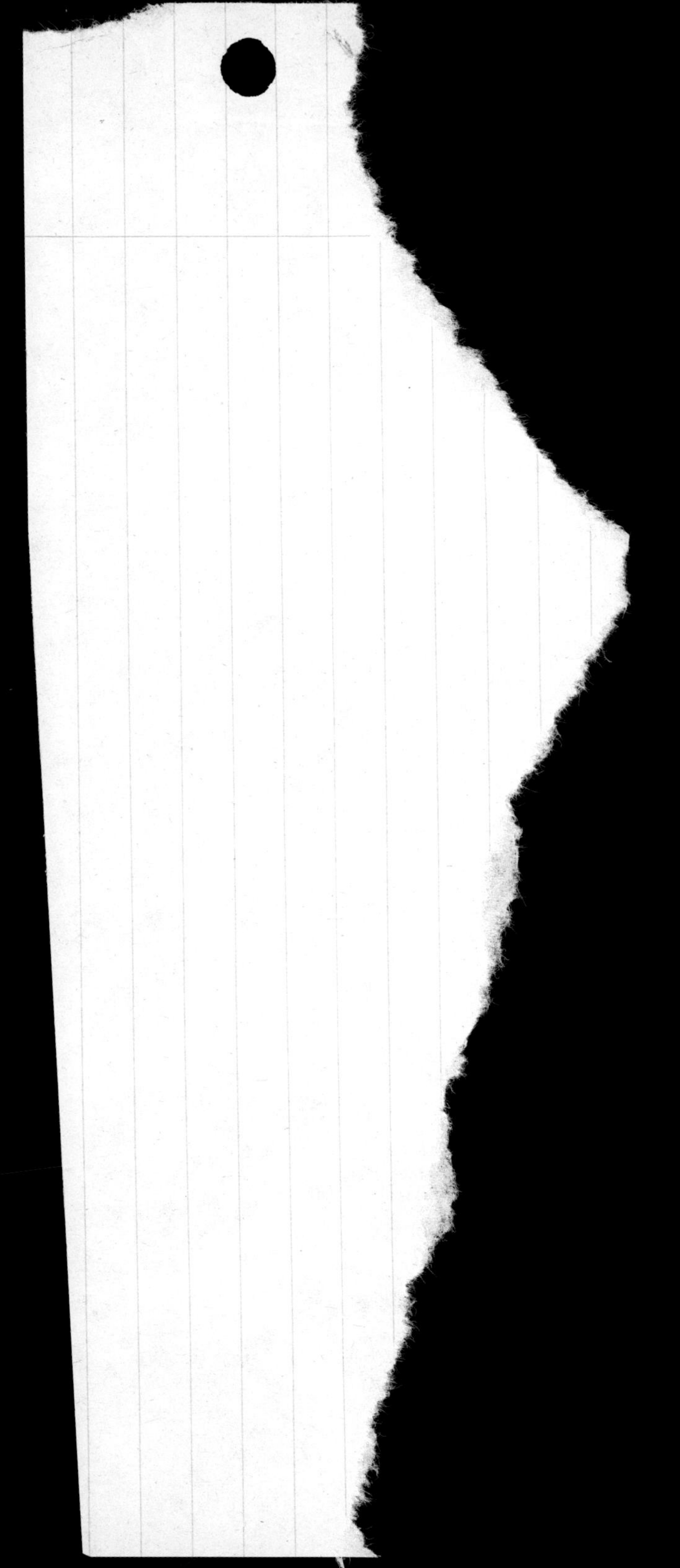

92¢ per bushel of corn, 45¢ for oats, 11¢ a pound for hogs, 35¢ for a dozen of eggs, and 62¢ per pound of butterfat. Prevailing prices in June, 1932 were: corn, 10¢ per bushel; oats, 11¢; hogs, 3¢ per pound; eggs, 22¢ per dozen; butterfat, 18¢ per pound.[30]

Throughout the summer months of 1932 evangelists of cost of production carried the farm strike message to county seats and crossroad hamlets across the length and breadth of the cornbelt. The *Iowa Union Farmer* announced on June 1 there were more Holiday meetings than the paper had space to report. Pledges of support trickled in from many Iowa counties. A county paper in the far northwest corner of the state noted, "Lyon County is hearing a lot about the farmers' holiday. The plan may or may not be workable, but it's morally right even if it isn't economically sound."[31] John Bosch, traveling alone, visited most counties in North Dakota and Minnesota. As momentum increased, South Dakota farmers met at Yankton July 20 to launch a state Holiday Association and a North Dakota organization began with a meeting at Jamestown, July 28. On July 29 a large farmer convention at St. Cloud listened to Milo Reno and E. E. Kennedy, then organized a Minnesota Farmers' Holiday Association and elected John Bosch president.[32] On July 26 a caravan of fifty automobiles with the Pierson community band in the lead toured towns and villages in Woodbury and Plymouth Counties and the following day at Pierson, Glen Miller addressed a crowd of 1,500, one of the largest audiences yet attracted in Iowa.[33] John Simpson defended the strike to the nation-wide audience of the National Farm and Home Hour: "The bankers are really on a strike in which they refuse to pay what they owe except on their own plans. Laborers for fifty years have used the strike successfully. I see no reason why farmers should not have the same rights as bankers and laborers. . . ."[34] Organizers borrowed the technique of the North Dakota Non-Partisan League two decades earlier for everywhere they traveled they asked farmers to sign pledges

---

[30] Lawrence, "The Farmers' Holiday Association in Iowa," 24-26; Saloutos and Hicks, 443.

[31] *Rock Rapids Reporter,* quoted in *Sioux City Journal,* August 4, 1932.

[32] *Willmar* (Minnesota) *Tribune,* August 1, 1932; *Sioux City Tribune,* July 20, 1932.

[33] *Unionist and Public Forum* (Sioux City), August 11, 1932.

[34] Quoted in *Iowa Union Farmer,* June 29, 1932.

to support the forthcoming withholding action. These enthusiasts had an inspired sense of numbers: they reported that half a million midwestern farmers had signed.[35]

Original plans called for the withholding movement to begin July 4 but the day passed without incident in the cornbelt. In the rapid build-up organization plans were not complete, but economic factors also may have been important. The declining trend in hog prices had momentarily reversed in June and the price rose to a high for the year of $5.15 per hundredweight on July 12. Cattle prices rose from $7.35 per hundredweight in mid-May to $8.60 on July 1. A few hopeful observers believed the low tide of depression had passed. However, a price slump on July 13 destroyed sanguine hopes and the price index began a steady descent that did not halt until December.[36] County chairmen of the Iowa Holiday Association met in Des Moines on July 30 and set August 15 as the starting date for withholding goods from market. Then on August 10, with considerable equivocation, the *Iowa Union Farmer* announced that the strike had really commenced August 8.[37]

The crusade to organize farmers to strike for cost of production had spread swiftly across the farmbelt. Depression had fired a resentment, like that which kindled the Cedar County Cow War, more virile than even the leaders suspected. The organizing drive galvanized forces independent of the Holiday Association that had not been reckoned with. What was designed as a peaceful withholding action became in its active phase a spontaneous and undisciplined movement, beyond the control of those who had conceived it and attempted to direct it.

---

[35] *Willmar Tribune,* August 2, 1932.

[36] Julius Korgan, "Farmers Picket the Depression" (unpublished Ph.D. dissertation, American University, 1961), 31-32; Soth, 22.

[37] *Sioux City Tribune,* July 30, 1932.

Three

A MOVEMENT LIKE WILDFIRE

This movement threatens to sweep the midwest like wildfire. It is a protest against an unbearable economic situation, a situation in which farmers can't even support families, let alone pay taxes and debts. This is a serious situation; it would be a mistake to minimize the dangers [Mayor W. D. Hayes of Sioux City, Iowa, *Sioux City Journal*, August 22, 1932].

The farm strike that began at Sioux City on August 11 was a different movement from that planned by the leaders of the Farmers' Holiday Association. In all the preceding build-up there had been no mention of picketing, yet at the very inception of the withholding movement farmers in Plymouth and Woodbury Counties patrolled highways and threatened non-cooperating farmers who tried to market their produce. Milo Reno conceded in a private letter, "I have not favored, at any time in our farm strike movement, picketing, because of the danger of loss of life and property, although I do feel that the action of the boys on the picket line has done more to focus the attention of the powers that be to the real facts of the situation than any other thing." [1] Usher L. Burdick, president of the North Dakota Farmers' Holiday Association explained, "The Holiday Association sprang up of its own accord. Some individuals were responsible for unifying a widespread national feeling, more pronounced in the farmbelt, into concerted action. Without any of the leaders in the field to-

[1] Milo Reno to Ben McCormack, November 8, 1933.

day the Holiday movement would have gone on."[2] The impulse provided by the Holiday leaders in northwest Iowa triggered a spontaneous grass-roots movement and prompted action from groups seeking redress of local grievances. The Farmers' Holiday Association was carried along on the unexpected flood tide.

The real purpose of the strike as Milo Reno conceived it, to force legislative action for cost of production, passed almost unnoted in press and periodical reports of events in the cornbelt. Almost universally the strike was interpreted as an attempt of farmers to hold their goods off the market until shortages forced up prices. This probably reflected the sentiments of most of the men who picketed the roadways. One reporter who lingered among the pickets heard Reno's name mentioned only once.[3] A striker told the Woodbury County attorney that the farmers aimed to achieve two things: first, to keep goods off the market until prices improved, and second, to demonstrate that farmers could hang together.[4] Motives for the disjointed and impetuous action on the highways of Iowa can never be determined; perhaps the best conclusion was that of a seasoned farm editor who noted, "After twelve years of this [legislative indifference] it relieves a farmer's feelings a good deal to throw a rock through a windshield or to take any positive step, no matter how futile it may ultimately prove to be, that seems to lead toward better prices."[5]

A farm strike was declared on August 8 and there was some sporadic activity around Des Moines, but the focal point quickly shifted to Sioux City where pickets appeared on the highway three days later. In the vanguard were 250 local milk producers for whom the strike was the culmination of grievances that had been accumulating for ten years.

Dairy farmers were the striking force of the depression farmers' protest. They were the first echelon in Iowa in 1932 and they

---

[2] *Farm Holiday News,* January, 1934. Burdick was subsequently member-at-large of the House of Representatives from North Dakota. He was the father of the present United States Senator from that state, Quentin Burdick.

[3] Josephine Herbst, "Feet in the Grass-roots," *Scribner's,* XCIII (January, 1933), 46-51.

[4] *Sioux City Journal,* August 19, 1932.

[5] Donald R. Murphy, "The Farmers Go On Strike," *New Republic,* LXXII (August 31, 1932), 67. Murphy succeeded Henry A. Wallace as editor of *Wallace's Farmer and Iowa Homestead.*

instigated both the Wisconsin Milk Strike and a spirited movement in New York State in 1933. Milk producers had a legitimate economic grievance: in Sioux City the farmer's price per pound of butter had declined from 40¢ in 1927 to 19¢ in 1932. The active role of dairy farmers, nevertheless, affirms the hypothesis that farmers most aggressive in protest were not those at the bottom of the economic ladder; the proportional decrease in income from dairy products had been less than that for any other single agricultural commodity.[6] Organizing dairy farmers was simpler than other producers. Since milk was highly perishable, a "milk-shed" area extended for no more than a twenty- to thirty-mile radius around a major urban center. Milk producers were relatively few in number and their total supply was sold to a handful of local distributors; price concessions could more surely be forced by a withholding action.

The J. R. Roberts Dairy Company, the largest in Sioux City, had gradually extended its milk-shed area, assuring a constant surplus of milk. This permitted the company to dictate its own price. In 1932, as milk production increased due to cheaper feed prices, the Roberts Company built a new plant, lowered the price paid producers and raised their retail price. A consumer paid 8¢ a quart for milk for which a farmer received 2¢. Twenty-eight aggrieved farmers organized a Producers' Cooperative Association in May, 1932 and chose I. W. Reck of Sioux City as chairman. During the summer the infant association signed up 900 members, a sizeable proportion of those who shipped milk to the Sioux City market. The refusal of J. R. Roberts to negotiate with the new association and his boasts that he had broken cooperatives in the past stiffened the determination of the leaders and forced them to turn to a withholding action as the only means of winning concessions.[7]

The campaign to form a milk producers' cooperative coincided with the organizing drive of the Farmers' Holiday Association. Many farmers shared membership in both groups. As the date approached for the Holiday withholding action, enthusiasts within

[6] U.S.D.A., *Yearbook of Agriculture, 1935*, 681.

[7] I. W. Reck, personal interview, March 12, 1962. Korgan, "Farmers Picket the Depression," 34 ff.; Sioux City Milk Producers Cooperative Association, *Record of Progress* (Sioux City, n.d.), 1, 3; *Unionist and Public Forum*, March 3, July 28, 1932.

the milk cooperative insisted, over the protests of Reck, that their organization should seize the opportunity to test their strength against the distributors. On August 10 the Producers' Association presented to the Sioux City distributors a demand for an increase in price from the present $1.00 per hundred pounds to $2.17. That evening, 250 cooperative members pledged their support to an immediate withholding movement and in order both to avoid public inconvenience and embarrass the distributors' retail trade, arrangements were made to distribute free milk in the city.[8]

On the night of August 11 fifteen trucks carrying milk to the city distributors were halted west of Sioux City and the milk of two of them was dumped on the pavement. By the succeeding evening farmers had extended their vigil to Cherokee and Plymouth Counties and roads to the north as well as the west of Sioux City were patrolled. Livestock trucks were also halted and the drive of milk producers coalesced with a general attempt to seal the Sioux City market.[9] On Saturday, August 14, 1,500 farmers stationed over five highways virtually blocked all shipments from the north and west. Pickets leaped on running boards as farm trucks crept up hill in low gear or threw logs and threshing machine belts in the path of approaching vehicles. Save for occasional fisticuffs and a few broken windshields, there was little violence; most trucks simply turned back. One hundred special deputies were recruited to keep open the roads, but the Woodbury County sheriff did not prevent the stopping of vehicles as long as techniques remained persuasive only. Receipts at Sioux City markets dwindled. The Producers' Cooperative claimed that 90 per cent of the customary supply of milk had been stopped and on Thursday, August 18 hog receipts were just half those of the preceding Thursday.[10]

As the selling holiday fanned out from Sioux City, the route from South Dakota was blocked and to the south farmers picketed the main highways leading to Onawa and Denison. To close the last breach in an increasingly solid barricade, a Holiday Asso-

---

[8] *Sioux City Journal,* August 10, 1932; I. W. Reck, personal interview, March 12, 1962.

[9] *Sioux City Journal,* August 12, 13, 1932; *Sioux City Tribune,* August 11, 1932.

[10] *Sioux City Journal,* August 14, 15, 19, 1932; *Sioux City Tribune,* August 15, 1932.

ciation was formed in Nebraska August 18. Five hundred farmers met at Dakota City and vowing that "no farm produce of any nature whatsoever" would pass their lines, marched out to take position on the main highways leading to Sioux City.[11] "This movement," said Mayor Hayes of Sioux City, "threatens to sweep the midwest like wildfire." [12]

Woodbury County strike leaders claimed that 90 per cent of farmers in the area either joined the picket lines or refused to sell produce. Had this statement been true there would have been no need for blockades to keep non-cooperating farmers from shipping milk and livestock to Sioux City.[13] Newspaper reports, although also probably exaggerated, permit a somewhat better estimate. The number of pickets at any one time or place was usually recorded as between 200 and 400. The largest number reported were the 1,500 who supposedly guarded the northern access to Sioux City on August 14. In 1930, 3,522 farm families lived in Woodbury County and 2,904 in Plymouth; allowing for such variables as non-farmers or outsiders on the picket lines, it may be concluded that those who participated in the farm strike constituted a sizeable minority of farmers in the two counties.[14] The milk producers' cooperative claimed only 900 members, the Woodbury County Farmers' Union numbered at best 250, the Plymouth County Union, 150. Therefore, active support extended beyond these nuclear organizations.[15]

Considering the unprecedented nature of the farm strike, public response was amazingly tolerant. The editorial attitude of the two leading metropolitan dailies in the area, the *Des Moines Register* and the *Omaha World-Herald*, was expressed in a *Register* editorial that declared, "It is the method of the holiday advocates, not the aspiration toward higher prices, that is dubious. . . ." [16] Of

[11] John L. Shover, "The Farm Holiday Movement in Nebraska," *Nebraska History*, XLIII (March, 1962), 57-58.

[12] *Sioux City Journal*, August 22, 1932.

[13] Lawrence, "The Farmers' Holiday Association in Iowa," 53.

[14] U.S., Department of Commerce, Bureau of the Census, *Fifteenth Census of the U.S., Population*, VI (Washington, 1933), 451-52.

[15] The estimate of Farmers' Union membership is based upon statistics in the *Iowa Union Farmer*, March 27, 1929, January 15, 1930, and January 13, 1932.

[16] Editorial, August 10, 1932; see also *Omaha World-Herald*, editorial, August 21, 1932.

twenty-two rural Iowa newspapers surveyed by the *Register,* few openly advocated support, but those entirely negative were fewer. Most agreed that farmers had a legitimate grievance but expressed fear of the forceful methods employed and doubt that self-action could achieve significant results.[17] At the storm center in Sioux City, the *Tribune* was unqualified in its endorsement while the larger paper, the *Journal,* shared the attitude of most Iowa newspapers: "Public opinion may not approve the farmers' strike . . . but it must admit that the tillers of the soil have sufficient provocation for definite action in some direction or other in an effort to survive." [18]

The American Federation of Labor and the Railway Brotherhoods sent resolutions of sympathy to the farm strikers,[19] but other farm organizations and leaders were hostile. Most judicious was the judgment of Henry A. Wallace, editor of the most influential farm journal in Iowa: "The farmers' strike is a mere gesture without economic significance except in the Sioux City territory. But gestures of this sort sometimes have far reaching significance, as for example, the Boston Tea Party . . . gradually they [the farmers] are beginning to wake up and it may be that this rather simple minded gesture known as the farmers' holiday will prove to have a greater significance than it is possible for economists or businessmen to understand." [20] Both the Grange and the Farm Bureau advised their members not to sign the withholding pledge or cooperate in any way with the strike. Edward A. O'Neal, president of the American Farm Bureau Federation, asserted that the farm strike "is sponsored by a limited group of misguided farmers with whom have become associated radical agitators." Milo Reno, quick to take offense when the Farm Bureau was involved, snapped back, "The statement of Edward O'Neal is exactly what might be expected from a man who has never undertaken a militant movement. . . . Mr. O'Neal's reference to radical agitators makes it appear that he has joined hands with Congressman Fish of New York in another 'red baiting exercise.'" [21]

---

[17] *Des Moines Register,* September 4, 1932; Lawrence, "The Farmers' Holiday Association in Iowa," 53.

[18] Editorial, August 16, 1932.

[19] *Sioux City Journal,* August 17, 18, 1932; *Des Moines Register,* August 30, 1932.

[20] *Des Moines Tribune,* August 19, 1932.

[21] Both quoted in *Sioux City Journal,* August 30, 1932.

In spite of the publicity attracted by the dramatic picketing of highways in northwest Iowa and the choking of farm shipments to the Sioux City market, the farm strike could achieve little. Two handicaps militated against its success: first, there was a lack of discipline and leadership and second, the economic objectives of the effort were unsound.

The first problem was the lack of discipline and control. Farmers' Holiday Association leaders like Milo Reno had not even anticipated anything so militant as picketing of highways; they were in no sense in command of activities around Sioux City and in no position to control them. Typical of the confusion was the action of farmers on the picket lines west of Sioux City who elected from their number officers for a Woodbury County Farmers' Holiday Association, unaware that officers had been appointed several weeks earlier by Reno.[22] Even local leaders had little authority to govern the actions of pickets scattered over highways for over a hundred miles.

The settlement of the milk strike on August 20 put the discipline of the Holiday movement to a severe test. J. R. Roberts, spokesman for the distributors agreed to an increase in price from the prevailing $1.00 per hundredweight to $1.80. Despite the settlement many pickets refused to allow milk trucks to pass. The Milk Producers' Cooperative assured the Holiday Association of its "loyal support in the future as in the past" but did not specify whether that support extended to the highways of Woodbury County.[23]

Increasing acts of violence marred the heretofore peaceful progress of the farm strike. On several roads pickets clashed with deputies attempting to escort trucks to market. In Nebraska an interstate freight train was halted, seals on cars broken, and cars uncoupled.[24] Most explosive, however, was the situation around Council Bluffs where Pottawattomie and Harrison County farmers blockaded Omaha from the Iowa side of the river. County officials, less compliant than those in Woodbury County, informed Clinton Savery, organizer of the blockade, that he would be per-

---

[22] W. C. Daniel, personal interview, March 14, 1962.

[23] *Des Moines Register,* August 20, 1932; *Sioux City Journal,* August 22, 23, 1932.

[24] *Sioux City Journal,* August 18, 22, 23, 1932; *New York Times,* August 24, 1932, p. 19.

sonally responsible for any violence or property damage. When Savery attempted to dissuade the pickets he was taunted with shouts of "sell-out!" "I have washed my hands of the entire mess. The strikers are beyond my control," he protested. On August 24 deputies in an automobile with tear gas cans mounted on the running boards confronted farmers armed with clubs and rocks. When forty-three were arrested, a sullen mob of 500, undaunted by machine guns in the hands of amateur deputies, swarmed over the courthouse lawn at Council Bluffs. Throughout a tense day they negotiated with law enforcement officials while buses waited at Des Moines, 150 miles away, to bring National Guard units. An eleventh hour parley and arranging of bail averted imminent bloodshed and tragedy.[25]

With the disorder at Council Bluffs, Nebraska authorities grew fearful of an expanding blockade around Omaha. Outside the city about a thousand men—tenants, farm boys, and city unemployed lacking shelter, food, or money—set up an impromptu camp. Not only were they without leaders but one informed a reporter they were there because they had "too much leadership already." As at Sioux City, the farm produce blockade coalesced with an attempt of milk producers to gain recognition from the city's dairies for a producers' cooperative marketing association and win higher prices for raw milk. Although the milk producers claimed to operate independently of the Farmers' Holiday, pickets halted dairy trucks and dumped milk at the city's outskirts. Governor Charles Bryan insisted that "the whole thing has been stirred up by agitators from Iowa" and ordered all pickets from Iowa arrested for inciting to riot. On the nights of August 30, August 31, and September 1, there were pitched battles of deputies and pickets at the Omaha city limits.[26]

A short-lived blockade around Des Moines constituted the only major disturbance outside the Missouri River Valley counties. Pickets appeared on the morning of August 27, blocking milk

[25] Shover, "The Farm Holiday Movement in Nebraska," *Nebraska History*, 59; *Sioux City Journal*, August 26, 1932; *New York Times*, August 26, 1932, p. 1; Governor Dan W. Turner to the writer, October 15, 1961. I am much indebted to former Governor Turner for submitting a twelve-page manuscript in response to my inquiries about his role and his interpretation of the Farmers' Holiday. This will be deposited in the library of the State University of Iowa.

[26] *Omaha World-Herald*, August 24, 30, 31, September 1, 1932.

shipments as there was no livestock market in the state capital. The Polk County milk strike was complicated, first, because unlike Sioux City or Omaha a producers' cooperative already existed here. The cooperative opposed the blockade and mounted riders armed with clubs on their trucks to escort them through the lines. Second, a group of "khaki shirt" veterans of the vanquished Washington bonus march flocked to the city limits to join the pickets. County officials, insisting pickets had abandoned peaceful methods, ordered the roads be kept open and demanded that all picket groups disperse. By August 30, nineteen farmers and five khaki shirts had been arrested.[27] This little flare-up coupled with the crises at Council Bluffs and Omaha demonstrated both that the farm strike was degenerating into violence and that law enforcement officials would tolerate little more interference with commerce on Iowa highways.

A rapid succession of critical events in the last days of August broke the back of the farm strike. Their patience exhausted, Woodbury County deputies swooped down on the picket lines the morning of August 26 and carried eighty-seven off to jail.[28] On August 30, fourteen pickets at Cherokee were injured by shots from a speeding auto. At Clinton, across the state, another angry mob defied armed deputies to converge on the county jail where three pickets were held prisoner.[29] Under duress of this sort, Reno and Chalmers proclaimed on September 1 a "temporary truce" in the farm strike effective immediately. An unsympathetic *New York Times* correspondent in Omaha observed, "[the] national leadership blew up, frightened at the appearance of the ugly monster into which its innocent child had so unexpectedly grown."[30]

The second handicap was the unsound economic foundation

---

[27] *Sioux City Journal,* August 30, 1932; *Des Moines Register,* August 27, 1932. Joe Gustavson, a participant in the Des Moines picketing, in an interview, August 1, 1961, gave little credence to press reports of the activity in Polk County. He declared that when khaki shirt veterans tried to join the lines they were rejected by the farmers. The blockade was so peaceful, he recalled, that when passenger cars stopped they were offered free milk, but trucks that tried to escape had to run through a maze of logs and hay bales.

[28] *Sioux City Journal,* August 26, 1932.

[29] *Ibid.,* August 31, 1932; *Des Moines Register,* September 1, 1932.

[30] September 4, 1932, Sec. 2, p. 7.

of the farm strike idea. From a practical standpoint, a farmer could not long participate in an embargo that deprived him of all income. There was livestock to be fed; there were families to be maintained. To keep marketable hogs and cattle on the farm meant added costs. Pressure of this sort led Farm Holiday leaders to announce on August 30 that the barricades would be relaxed sufficiently to allow hard-pressed farmers to market products.[31]

The peaceful withholding movement, the major objective of the Farmers' Holiday Association, had been lost sight of in the flurry of dramatic activity around Sioux City. Contradicting the exuberant claims of its organizers, it had been a failure. Receipts of livestock had decreased only at Sioux City; there was an increase at neighboring markets. Indeed, prices for farm products dropped to a year's low while the farm strike was in progress.[32] Pickets on Woodbury County highways had won wide publicity for the farmers' plight, but there was no conceivable possibility that blockading a single market could reduce the overall supply of agricultural products sufficiently to increase prices. Only the milk producers, who operated tangentially to the Holiday Association, could claim any concrete benefits from the farm strike.

In an unexpected sense, the farm strike had achieved better results than the original promoters could have expected. The spontaneous movement element that seized the initiative from the Holiday leaders in northwest Iowa publicized the farmers' plight and prompted political response more effectively than any ill-organized peaceful withholding movement. Elected political officials in the Midwest could not long remain apathetic confronted with such a manifestation of rural discontent. Governor Warren Green of South Dakota supported the farmers' plea for cost of production and hinted that he might invoke legal sanction for an orderly farm embargo. The Farmer-Labor Governor of Minnesota, Floyd B. Olson, was even more outspoken. "I am in sympathy with the strikers," he declared and in a rash moment stated he might invoke martial law to prevent marketing of farm

---

[31] *Sioux City Journal*, August 31, 1932.

[32] The number of hogs marketed in Iowa was 750,525 in July, 804,335 in August, 787,353 in September, 806,035 in October. Computed on the 1910-14 base, the index numbers for Iowa hog prices were: July, 58; August, 53; September, 49; October, 41; November, 38; December, 33. Soth, 22, 41.

products at less than cost of production.[33] Dan W. Turner of Iowa ventured no support for the strike, but he spurned requests that the National Guard be dispatched to the troubled areas. Thirty years later he avowed that he had always favored cost of production for farmers.[34] At Green's suggestion the governors agreed to meet in Sioux City on September 9 to hear appeals of the Holiday leaders and perhaps agree on a program of common action. The summoning of the conference provided a rationale for the "temporary truce" called September 1 by Reno and Chalmers.

The most outspoken dissident among farmbelt governors was Charles W. Bryan of Nebraska. Despite his long associations with farmer causes, the brother of the "Great Commoner" could see nothing but "hard feelings and some bloodshed" resulting from the Farmers' Holiday. Governor Bryan insisted that the farm problem was national in scope and could not be assuaged by local picketing and farmers' strikes. The real enemy, he argued, was "the powers in Washington" and the remedy was "the repeal of legislation which caused this condition and not temporizing with the effects." As for the governors' conference, he declared that "the people of this nation have suffered more as a result of surveys and conferences than any other alibis they have been afflicted with." [35] He did not attend the Sioux City meeting.

After watching a crowd of farmers estimated at 5,000 march through the streets of Sioux City, the four governors present [36] heard the testimony of forty witnesses. Milo Reno, speaking for

---

[33] *Sioux City Journal,* August 28, 1932; *Willmar Tribune,* August 22, 1932. Vince Day, private secretary to Governor Olson, criticized the Governor's off-the-cuff remark about an embargo. Day to Olson, August 26, 1932, Vince Day Papers, Minnesota State Historical Society, St. Paul. See also George H. Mayer, *The Political Career of Floyd B. Olson* (Minneapolis, 1951), 104-6.

[34] Governor Turner to the writer, October 15, 1961. In the late fifties, despite his more than seventy-five years, Turner was active in organizing the National Farmers' Organization.

[35] *Sioux City Journal,* August 28, September 2, 1932. The legislation to which Governor Bryan referred was "high tariff, high finance, the Esch-Cummins act and other legislation which gave one large business group special legislative advantages." *Des Moines Register,* August 28, 1932.

[36] Present at the conference were Governors Turner of Iowa, Green of South Dakota, Olson of Minnesota and Shafer of North Dakota. Also attending were representatives of the governors of Nebraska, Ohio, Wisconsin, Wyoming, and Oklahoma.

the Holiday Association, proposed a four-point program: (1) state mortgage moratoriums putting a temporary stop to all foreclosure proceedings; (2) a special session of Congress to enact the Frazier inflationary farm credit bill; (3) voluntary action by farmers to withhold goods from market; (4) most important and most controversial, a demand for state-enforced embargoes against the sale of farm products at less than cost of production. These proposals represented an advance over the simple and immediate objectives stated at the Holiday organizing convention in May. The request for a mortgage moratorium represented a new emphasis. The proposal for an embargo was also an innovation, appearing only this one instance in any Farmers' Holiday resolutions. It was a direct response by the Holiday leaders to the situation in northwest Iowa and was an attempt to gain legal sanction for the objectives those farmers had tried to achieve by picketing highways.

The governors expressed little sympathy for the far-reaching requests of the Holiday leaders. The embargo proposal would have compelled all farmers, regardless of their sympathy for the withholding movement, to keep produce at home. "When you insist on an embargo on farm products and picketing of roads, you ask the impossible," Governor Turner told Reno.[37]

The upshot of the governors' conference, therefore, was a series of tame resolutions to President Hoover including such time-worn panaceas as tariff protection for farmers, currency expansion, and a request that federal and private agencies desist from foreclosures. On the one proposal where the governors might have taken an initiative, state mortgage moratoriums, their memorial was silent. Although there was little in the governors' resolutions that resembled Farmers' Holiday demands, Reno stated that he was "on the whole" satisfied, but was disappointed by the lack of an embargo. This satisfaction was not shared by all Holiday members. John Bosch of the Minnesota Association declared that only Olson had supported the farmers' demands and called for a renewal of the strike until the governors changed their attitude. A few days later, the first convention of the Nebraska Farmers' Holiday Association resolved: ". . . we consider the governors' conference at Sioux City a dismal failure, and we call upon the

[37] *Sioux City Journal,* September 10, 11, 1932.

farmers, farm workers, and those dependent on Nebraska farming to join in the struggle for immediate action." [38]

The days following the governors' conference were a time of confusion and ambiguity for the Farmers' Holiday. Underscoring the lack of discipline in the movement, pickets remained on the highways despite the request of the national officers for a "temporary truce." On the eve of the Sioux City meeting, pickets threw bricks at deputies escorting trucks and injured the sheriff of Plymouth County. Four days after the conference a farmers' meeting at the Golden Slipper Dance Hall, just south of Sioux City, voted 436-249 to resume picketing; even W. C. Daniel, president of the Woodbury County Association, could not dissuade them. Picketing ceased in all parts of Iowa, but at two strong centers, one at James, north of Sioux City, and another at Correctionville, ten miles to the west, the roads were still blocked. "The situation is as bad as ever," the Woodbury County attorney complained and he joined the sheriff and the mayor in appealing for National Guard units. Governor Turner hesitated to dispatch troops and appealed personally to the strike leaders in Woodbury County. Turner's subsequent claim that his action terminated the "wildcat" picketing is doubtful. What did stop it is uncertain. Perhaps it was cold weather, the advent of "corn-pickin'" season, maybe the sheer futility of the action. Perhaps it was the threat of a farm leader who declared that a few radicals now dominated the picket lines and if the blockade continued "the very farmers that were picketing the roads and which have kept such splendid control of the sitution . . . will be out on the highways to see that the roads are kept open." [39]

Meanwhile, national Holiday leaders hastily evolved a plan less dangerous than blockading highways to maintain the momentum of the movement. An embargo "beginning at the farm gate" would be inaugurated on September 20; it would apply to grain and livestock only and participants were to refrain from picketing. The strength of the Holiday Association was not in the voluntary support of any wide segment of the farm population. This renewed withholding action was stillborn; commodity

[38] *Ibid.*, September 16, 24, 1932; *Willmar Tribune*, September 13, 1932.

[39] *Sioux City Tribune*, September 1, 1932; *Des Moines Register*, September 8, 1932; *Sioux City Journal*, September 8, 16, 18, 1932; *New York Times*, September 16, 1932, p. 5; Governor Turner to the writer, October 15, 1961.

receipts in Iowa were unaffected and even the *Iowa Union Farmer,* the voice of the Holiday Association, made no report on its progress.[40]

While picketing dwindled and the withholding movement languished in Iowa, the focus shifted northward where the Minnesota Farmers' Holiday Association launched a delayed offensive. The Minnesota organization pursued a course largely independent of the Iowa group. The declared purpose of its withholding action was to keep goods off the market until cost of production was obtained through *state and national legislation.* The movement was designed as a first step toward a plan for voluntary surplus control through which farmers would hold excess on their own farms. Credit for this more sophisticated approach belongs to John Bosch of Willmar, the young farmer who served as Minnesota state president. Bosch had been reared in a Populist family with a strong intellectual tradition and he took pride in his abilities in economic analysis. His brother and advisor, Richard, was a graduate of Commonwealth College and had studied economics as a graduate student with John R. Commons at the University of Wisconsin. The Bosches were close friends of Governor Olson and the Minnesota Holiday Association enjoyed the advantage, denied in Iowa, of close liaison with the state government.[41]

While Iowa farmers patrolled roads, Minnesota steadily continued to build its organization. When withholding began on September 21 it was claimed that fifty counties had been organized. Like the Iowa leaders, Bosch counseled against violence. "We won't picket, we'll just sit tight," he said on August 29 and two days later he announced that Minnesota farmers would use "Gandhi-like" methods.[42]

Nevertheless, as in Iowa, admonitions of the leaders were not entirely successful. Before the deadline date, farmers in Nobles County on the Iowa border blocked all shipments southward to Sioux City. No sooner had non-marketing begun than picket lines appeared in Lac Qui Parle, Chippewa, and Yellow Medicine

---

[40] *Iowa Union Farmer,* September 21, 1932.

[41] *Willmar Tribune,* August 1, 30, 1932; John Bosch, personal interview, April 1, 1962.

[42] *Willmar Tribune,* August 29, September 1, 1932.

Counties. In the latter county, a farmer who resented pickets trespassing on his property shot and killed Nordahl Peterson, a twenty-five-year-old striker. Picketing in Minnesota was more sporadic than in Iowa and fewer individuals participated. In only two spots was it maintained more than two days and the largest number involved was 500 at Anoka on October 12. Holiday activity centered in the southwest corner of the state where Minnesota joins Iowa and South Dakota. Outside this area, there was a short-lived attempt to halt shipments from the north to the giant livestock marketing center at South St. Paul. The Minnesota Holiday Association was true to its word; the withholding action and picketing ceased October 22, just thirty days after it began.[43]

A crucial national political campaign was in progress in the autumn of 1932 and candidates for national office could no more remain apathetic to symptoms of disaffection in the farm states than could midwestern governors. President Hoover, "alarmed by the growing farm revolt in the West," discussed with advisors on September 26 a plan to raise commodity prices by selling fifteen million bushels of wheat to China. Two days later the President informed Governor Turner he had promoted a "preliminary discussion" among eastern mortgage concerns and government agencies on the question of farm mortgages.[44] Nevertheless, the reception the state capital gave to President Hoover on October 4 presaged what his home state of Iowa had in store for him in the election one month hence. Despite fears that some Holiday farmers might dispose of a small portion of their surplus by flinging it at the President of the United States, the crowd was orderly and the presence of the Holiday was evident only from several trucks that paraded the streets of Des Moines bearing slogans.[45]

Governor Roosevelt, speaking at Sioux City on September 29, condemned the "spend-thrift" policy of the Hoover administration

---

[43] Details of Holiday activity in Minnesota are reported in detail in *Willmar Tribune,* September 21–October 22, 1932 and *Minneapolis Journal,* September 21-24, 1932.

[44] *Farmers' Union Herald,* October, 1932; *New York Times,* September 16, p. 15; September 29, p. 1; September 30, 1932, p. 2.

[45] *New York Times,* October 1, p. 3; October 5, 1932, p. 19.

and criticized the Smoot-Hawley tariff, but to the disappointment of some of his 25,000 listeners, he outlined no specific program for agricultural relief. Nevertheless, leaders of the Farmers' Union and Holiday Association were enthusiastic supporters of the Democratic candidate. John Simpson had toured the Midwest, combining pleas for support of the Farmers' Holiday with campaigning for Roosevelt. The *Iowa Union Farmer* endorsed him. There was good reason for the enthusiasm of the radical farm leaders. The brief Democratic national platform had promised farmers "prices in excess of cost"; Farmers' Union zealots interpreted this as meaning "cost of production." [46] At Sioux City, Glen Miller of Iowa, John Simpson, and other Union leaders talked with Roosevelt. They alleged he had promised that if elected he would devote more time to agriculture than any other single problem, that he would press for legislation refinancing loans at a low rate of interest and guaranteeing to farmers cost of production prices.[47]

Only a minority of Iowa farmers might join picket lines, but in overwhelming majority they repudiated the agricultural policies of the Republican administration. Hoover carried only six of the Iowa counties (all in urban areas) and in northwest Iowa, traditionally Republican territory, Roosevelt's majority was better than two to one.[48] Of the governors who had attended the Sioux City conference, only Floyd B. Olson, a Farmer-Laborite, survived. Iowa and South Dakota elected Democrats and in North Dakota, the Non-Partisan League returned to power with the election of William L. Langer.

Milo Reno and Holiday leaders had every reason to be hopeful that the new Democratic administration would soon produce agricultural legislation that would satisfy their demands. Shortly following the election, all state Holiday presidents met in Omaha and declared that any general strike or withholding movement would be suspended until the incoming administration had the opportunity to redeem what they considered two basic pledges:

---

[46] The official Farmers' Union biography of Milo Reno in 1941 still quoted the 1932 Democratic platform as containing the words "cost of production." White, 86. See *infra*. ch. 6.

[47] *Iowa Union Farmer*, October 5, 1932.

[48] Edgar Eugene Robinson, *They Voted for Roosevelt* (Stanford, 1947), 89-93.

refinancing farm mortgages at a low rate of interest and stabilizing farm prices at a cost of production level.[49]

The Farmers' Holiday Association in just six months had mustered a political strength that could command the attention of governors and win promises from presidential candidates. The irony was that the strength of the movement was a tempestuous and little organized force whose allegiance to the Holiday Association was tangential. The peaceful withholding movement that the organization had planned had been a total failure. Irrational picketing of highways and mob-like courthouse demonstrations not condoned by the leadership had focused nation-wide publicity on the farmers' plight, prompted political appeasement, and provided bargaining strength for the Farmers' Holiday Association.

---

[49] Milo Reno to B. H. Aldridge, February 19, 1933.

Four

WE MUST LEAD THEM

We cannot command them to forget about prices and fight for something else. We must lead them, or fascism will—and against us [Harrison George, "Causes and Meaning of the Farmers' Strike and Our Tasks as Communists," *The Communist,* X (October, 1932), 931].

Depression weakens the moderation and patience that are the moorings of any stable democratic political system and heightens the appeal of extremist political movements. The early thirties abounded in messiahs who preached salvation through technocracy, anarchy, inflation, fascism, or any of a myriad of panaceas. A few advanced carefully reasoned programs and claimed a small, but devoted coterie of supporters. Most were crackpot, and if they commanded any following at all, it was because of the magnetism of some dominant leader.

The troubled farmlands were a fertile breeding ground for extremist movements. The spontaneous, leaderless insurgency that erupted under the title of Farmers' Holiday circumscribed a wide variety of genuine grievances and emotional prejudices. Protesting farmers had demonstrated their dissatisfaction with slow and orderly political processes; they shared traditional agrarian hostility to "Wall Street" and the "bureaucrats"; they looked to simple and immediate measures to remedy a complex situation. Within this framework either a leftist or rightist program could be constructed. Part of the history of agrarian protest in the thirties revolves around an internal struggle for leadership between a Populist-oriented capitalist revitalization group and a radical left wing faction that demanded more sweeping changes. This ambivalence is illustrated by the fact that the two external movements that made inroads into areas affected by the Farmers' Holiday were almost antithetical in their ultimate objective. One, the Modern Seventy-Sixers, sought to revitalize the free enterprise system by breaking, by force if necessary, the stranglehold of corporate capitalism. The other was the Communist party of the United States. Both flourished in the identical Missouri River counties and reached the apex of strength at about the same time. Available evidence indicates that both drew their supporters from the ranks of family farmers with heavy mortgage burdens or tenants, many of whom were former owners.[1]

Ambiguity of this nature is common in agrarian protest movements. Populism gave rise to anti-semitism and native fascist sentiments that have recently attracted the attention of historians, but out of Populism emerged, too, the grass-roots socialism of Julius Wayland and the *Appeal to Reason.* The Cooperative Commonwealth Federation of Saskatchewan, a socialist organization, joined forces for a two-year period with the Social Credit party of Alberta which sought to preserve capitalism through reforming the monetary system.[2]

The point is not that certain types of farmers in relatively industrialized countries are potential fascists or Communists, but that they have a certain propensity to radicalization under conditions of acute distress. When such radicalization will eventuate and which way it will turn, the analyst of social stratification is not in a position to predict. His knowledge does enable him to estimate the relative chances for such a development, but only in the sense that certain types of farmers are more likely to be affected than others. Obviously, local conditions, historical antecedents, the acuteness of the crises, and the intensity of the organizational drive on the part of the totalitarian movement will play a role and can be judged only in specific cases.[3]

---

[1] For a discussion of the incidence of present or former property holders in the Farmers' Holiday, see *supra,* p. 9. Communists were surprised and distressed when a questionnaire circulated among 408 delegates to their second Farmers' National Relief Conference at Chicago in 1933 revealed 74.8% to be owners and 25.2% to be tenants. George Anstrom, "Class Composition of the Farmers' Second National Conference, Chicago, 1933," *The Communist,* XIII (January, 1934), 47-52. See *infra,* ch. 11.

[2] Lipset, 123.

[3] Reinhard Bendix, "Social Stratification and Political Power," *Class, Status and Power,* eds. Reinhard Bendix and Seymour M. Lipset (Glencoe, 1956), 602.

The most amazing fact about the Modern Seventy-Sixers is that such an organization existed at all and attracted some serious supporters. Its career was brief, its membership small, and it might be dismissed as some quaint political aberration but for the fact that included among its members were all the key local Farmers' Holiday leaders from northwest Iowa and that some of its supporters were involved in the most violent episodes the direct action movement provoked. The short flurry of success this organization enjoyed underscores the susceptibility of desperate farmers to the blandishments of demagogues and political panaceas.

Lester P. Barlow of Stamford, Connecticut arrived in Sioux City in October, 1932, bearing, according to his account, a commission from the Democratic National Committee to campaign in the area for Franklin D. Roosevelt. He claimed to speak as a political independent, endorsing only Roosevelt, not the entire Democratic slate. This was no new role for Barlow. He had a reputation as a forceful stump speaker in his native Stamford and ten years earlier he had crusaded in Iowa for the Non-Partisan League. He campaigned for LaFollette in Cleveland, Ohio in 1924, the only city captured by the unsuccessful Progressives. He was an outspoken and dramatic platform performer with an uncanny ability to sway large crowds of people.[4]

Barlow had gathered together a potpourri of personal recollections and suggestions for radical social change in a book, *What Would Lincoln Do?*, published the preceding year. Initiative was stifled and progress chained, he contended, by a willful conspiracy of gangster politicians and racketeers of "high finance" who had led the United States into war, engineered the Dawes and Young Plan to save huge investments abroad, and reduced American agriculture to ruin. He would have this power shattered by the mobilization of the common citizenry into a revived national Non-Partisan League. Clustered in military-like companies to be known as the "twenty and one," they would infiltrate the major parties to nominate and elect candidates committed to a bold reform program. As an immediate step to curb unemployment the federal government should embark on a plan to build a network of motor expressways across the nation—arresting

[4] Lester Barlow, personal interview, January 4, 1962. *Unionist and Public Forum,* October 20, 1932.

depression, congestion, and traffic fatalities in one inclusive step. A constitutional amendment declaring $500,000 to be the maximum total wealth any individual might own would inaugurate a "limited capitalism" where gambling on the stock and commodity markets would disappear, the profit incentive would be eliminated, and wealthy parasites put to work.[5]

Lester Barlow was a fantastic figure who by a combination of incredible rashness, herculean ego, and genuine mechanical genius lingered for more than two decades on the threshold of political influence. He accumulated a moderate wealth as an inventor and mechanical engineer. He had served in Mexico in 1914 as a staff officer in Pancho Villa's army. He invented a depth bomb used by the navy in World War I and twenty years later collected from the United States Government $600,000 for infringement of his patent rights.[6] In 1932 he claimed to have invented a liquid oxygen bomb so powerful that no nation would ever again dare resort to war. The proposal won for him an interview with President Hoover and a friendly response when he carried his plans to Moscow and Berlin.[7] Barlow's motor expressway was the subject of the Phipps-Robsion resolution passed by the New York State Senate in 1929 and the proposal was endorsed by Governor Franklin D. Roosevelt.[8] As president-elect, Roosevelt took Barlow with sufficient seriousness to confer with him during the inter-regnum on both his highway and armament proposals.[9]

Barlow's successes came not from any distinctive ideology, but from the charisma of a compelling leader in an area and movement where little leadership had been exercised. In a whirlwind tour through northwest Iowa he spoke at Sioux City, Correctionville, Lemars, Clear Lake, and Onawa. When Vice-President Charles Curtis spoke in Sioux City, Barlow, with typical dramatic flare, offered to pay him $100 if he would respond to ten ques-

[5] (Stamford, Connecticut, 1931), 56-138, 159-75.

[6] *New York Times,* August 28, 1940, p. 8; March 26, 1955, p. 23; Barlow, 35-36.

[7] *New York Times,* May 6, p. 9; August 13, 1932, p. 9.

[8] Barlow, 92, 97; *New York Times,* January 28, 1931, p. 4; Barlow to Roosevelt, April 4, 1930; Roosevelt to Barlow, April 5, 1930. Roosevelt Papers, Hyde Park, New York.

[9] M. A. Lehand to Barlow, December 2, 1932. Roosevelt Papers.

tions propounded by Barlow.[10] His campaigning served as much the purpose of advancing his Modern Seventy-Sixers as it did the candidacy of Roosevelt. On October 19 Barlow presided at an organizing meeting which called together major Farmers' Holiday leaders from northwest Iowa. The committee to organize the Modern Seventy-Sixers included W. C. Daniel, president of the Woodbury County Farmers' Holiday; C. J. Schultz, who held the same position in Plymouth County; Joe Trudeau, president in Union County, South Dakota; and Fred Kriege, president in Dakota County, Nebraska.[11] At a mass meeting on October 26, Milo Reno shared the platform with Lester Barlow and extended his cordial support to the Modern Seventy-Sixers.[12] Barlow's activities were given laudatory coverage by the pro-Holiday *Sioux City Tribune* and Wallace Short, former mayor of Sioux City and editor of the *Unionist and Public Forum,* heartily endorsed Barlow's idea of waging war on economic corruption: "The Modern 76ers look like one good way to do it. I have not heard anyone proposing a better way and we had better do something and do it now." [13]

Barlow, meanwhile, had supplemented the broad social pro-

---

[10] *Sioux City Tribune,* October 19, 1932.

[11] *Ibid.,* October 20, 1932.

[12] *Ibid.,* October 26, 1932. In answer to an inquiry concerning the Seventy-Sixers five months later, Reno wrote, "This organization, while not an auxiliary of the Holiday Association, is, at least, a movement in the right direction." Reno to W. L. Achenbach, March 1, 1933.

[13] *Sioux City Tribune,* October 26, 1932. Short, a close friend of Reno, was alleged to be the hidden power behind the Holiday movement in northwest Iowa. Short began his career as a liberal minister; his exemplar was the distinguished Rev. Washington Gladden whom he described in 1937 as the "greatest man I ever knew." He lost his pulpit at the First Congregational Church of Sioux City because he refused to support prohibition. Elected mayor of Sioux City on a Labor ticket in 1918, his tolerance toward the maligned Industrial Workers of the World led his opponents to designate the city "Wobblies paradise" during his tenure of office. Despite attempts at recall, he served three terms and represented Woodbury County in the Iowa Assembly between 1930 and 1932. Wallace Short founded the *Unionist and Public Forum* in 1927. In the mid-thirties he was a strong supporter both of Father Charles Coughlin and Huey Long. Still active in 1948, he was the first Iowan to endorse the presidential candidacy of Henry A. Wallace. His career still awaits the attention of historians. The only biography is an affectionate and inadequate memoir, Mrs. Wallace Short, *Just One American* (Sioux City, 1943). His statement about Gladden is in *Unionist and Public Forum,* July 17, 1937. I am indebted to Mr. Fred Stover of Des Moines for additional information.

gram outlined in *What Would Lincoln Do?* He called for the issuance of "76er script" to suffice temporarily for currency in northwest Iowa. More important, obsessed as he was by conspiracy and violence, he predicted that a forceful revolution was impending in America; the solid citizenry of the rural Midwest must mobilize to prevent it. In one rash moment, Barlow offered to place himself at the head of a citizens' army of 25,000 to be headquartered in Sioux City. Once organized, Barlow would submit to the federal government a nation-saving program which would include government ownership of all utilities, operation of all industries by the citizens' army "until a new order may be established," elimination of the stock market, and revision of the structure of representative government.[14]

Barlow's objectives and method shared much in common with such co-existent native fascist organizations as the Silver Shirts of William Dudley Pelley.[15] Nevertheless, Barlow insisted that the Modern Seventy-Sixers, like the Non-Partisan League, aimed at "revolution through the ballot-box." The citizens' army was designed to preserve democratic methods; to prevent, not foment, revolution. One is reminded of the dictum of Georgi Dimitrov of the Comintern: "It is a peculiarity of the development of American fascism that at the present time it appears principally in the guise of an opposition to fascism." [16] Nevertheless, the term

---

[14] Barlow to Fred Kriege, May 1, 1933. J. Fred Kriege Papers, private collection of Mr. John Kriege, Hayward, California.

[15] Morris Schonbach, "Native Fascism During the 1930's and 1940's: A Study of Its Roots, Its Growth, and Its Decline" (unpublished Ph.D. dissertation, University of California, Los Angeles, 1958), 303-16. The program of the Seventy-Sixers is almost precisely in accord with the elements of fascism in the American context as defined by Victor C. Ferkiss, "Populist Influences on American Fascism," *Western Political Quarterly,* X (June, 1957), 350-73.

[16] William Z. Foster, *History of the Communist Party of the United States* (New York, 1952), 321. Mr. Barlow, a quick-witted man of 73, was not troubled by my suggestion that the Seventy-Sixers might have fascistic overtones. He denied this because, first, he always stressed political, not forceful methods; second, a major point in his program was free speech even for his most powerful opponents; third; he was not anti-semitic and he had refused to support Father Coughlin on these grounds; fourth, while he was no sympathizer with Communism he was no "red-baiter"; he viewed Soviet progress with satisfaction at the time of his 1932 visit. His proposal for a citizens' army, which he conceded was rash, could be understood only in the context of the situation in early 1933, he argued, when he honestly believed some attempt would be made by reactionary interests to destroy

"fascist" applied to the Modern Seventy-Sixers is a label, not a serious analytical tool. Fascist movements arise from economic and psychological frustration, but all movements of frustration are not necessarily fascistic. The term bears a European connotation and a few parallels between incipient American movements and the full-blown specter of powerful European movements are not adequate to classify both in the same genre. Only a careful study comparing social movements in nations that succumbed to fascism and those that did not, a study that would end, not begin, by defining terms, would be sufficient to establish the case.[17]

Barlow left Sioux City in early November advising, "It [the Seventy-Sixers] should continue until after the election of 1934 and then disband before it has a chance to become rotten." [18] Initiative for pushing the organization was assumed by the *Unionist and Public Forum* and meetings and activities were reported during the six subsequent months. The strongest unit was organized in O'Brien County, north of Sioux City, where the Seventy-Sixers functioned in place of the Farmers' Holiday As-

---

representative government. Barlow viewed with disinterest and disdain the charges of conspiracy and the attacks upon elitism by Senator Joseph McCarthy in the fifties. Some of the difficulties in attempting to classify a movement such as Barlow's are discussed in Schonbach, "Native Fascism During the 1930's and 1940's," 18-38, 263.

[17] A study that might contribute to such an analysis is Charles P. Loomis and J. Allan Beegle, "The Spread of German Nazism in Rural Areas," *American Sociological Review,* XI (December, 1946), 724-34. The main emphasis is upon the appeals of Nazism in selected farming areas in Schleswig-Holstein and Bavaria, but the authors draw undocumented comparisons to agrarian movements in the United States. They found, for example, that electoral support for the Nazi party was greatest among farmers whose basic orientation was to small-scale primary groups and whose experience with large-scale bureaucratic affiliations was insignificant. The Nazis won strongest support in regions where family-sized farming predominated, where hogs and cattle were principal products, and where mortgage foreclosures were relatively frequent. The authors note the presence in these areas of antiforeclosure demonstrations. A major drawback to such a comparison with the Farmers' Holiday Association is the fact that in the United States Communist farm organizations drew their support from the same type of regions and the identical constituency. Given the diffuse nature of the depression farmers' movement in the United States such variables as historical antecedents, the acuteness of the crisis, and particularly the intensity of the organizing drive on the part of the totalitarian group determined the ideological direction this essentially economic movement would take.

[18] *Sioux City Tribune,* October 26, 1932.

sociation. Leaders there and in Plymouth County were arrested when martial law was declared in northwest Iowa in May, 1933 and the state took sufficient interest to investigate the alleged subversive activities of the Seventy-Sixers. Barlow continued to exhort his followers from afar. He wrote Short, "Roosevelt can't beat the gangsters unless we back him with a club and with our chin out," and in a letter to Reno he threatened, "If Roosevelt don't [sic] have a 'New Deal' we will be on the march in March." [19] (Barlow became disillusioned with the new president and in May he described him as an "irresponsible lightweight, a political adventurer and an opportunist.")[20] He was not distressed that his Sioux City followers produced no citizens' army. By 1934 he had found a new friend and political patron in Senator Huey Long. Apparently he enjoyed a close liaison, for Barlow alleges that shortly before his death Long proposed that they join for a speaking tour across New England and the Midwest. Thirty years later, Barlow nostalgically recalled, ". . . if the Honorable Huey P. Long had not been murdered by the politically scared, the history of this nation would have been much different than it is today." [21]

The Modern Seventy-Sixers was an abortive movement; Barlow believed the Sioux City contingent numbered about 200 and that a few hundred more may have been added following his departure. The movement functioned in a leadership vacuum and attracted drifting and confused men stricken by anxiety and frustration. The involvement of local Holiday leaders and the friendship of Reno indicate how closely the farm protest skirted the bounds of lunacy and courted totalitarianism. Had there been another year of political inaction, had there been no legislation to restore agriculture—what then?

The American Communist party observed the rural crisis with increasing interest and determined that aggrieved farmers should

---

[19] *Unionist and Public Forum,* February 2, 1933; Barlow to Reno, February 7, 1933.

[20] Barlow to Kriege, May 1, 1933. Kriege Papers.

[21] Lester Barlow to the writer, October 19, 1961; Barlow, personal interview, January 4, 1962. Barlow claimed that Long's "Share-the-Wealth" plan borrowed from his own plan for "limited capitalism." Considering that each time I have been able to check the veracity of Barlow's claims they have proven true, I no longer dismiss them lightly.

be incorporated into a proletarian "united front" along with the unemployed in cities. Studies of the Communist party have either ignored or dismissed this attempt to infiltrate the depression farm protest.[22] This leaves unexplored a chapter in the history of American radicalism that is revealing both as to the strategy and tactics of the Communist party and the nature of the farmers' movement. The Communist party assigned some of its most capable organizers to the agrarian program, and the program prompted theoretical debates that struck at the core of party operation and practice. American Communists had first taken cognizance of the farmers' increasingly critical situation in 1929 when the party's theoretical journal began a series of articles on the agrarian issue. The recurrent theme was the existence of a "scissors phenomenon," the effect of continuing economic crisis to squeeze out moderate-sized farmers and widen the gap between prosperous "factory farmers" and impoverished small operators. The articles postulated that class lines were hardening in the American countryside.[23]

The most important manifestation of this newly found interest was a comprehensive plan of action referred to as the "draft program" and prepared by the Agricultural Committee of the party. Entitled "U.S. Agriculture and the Tasks of the Communist Party, U.S.A.," it was considered at the Seventh Party Convention in 1930 and published in advance.[24] The party had neglected the agrarian masses, the report began, but now "tasks have accumulated, which, together with developing opportunities, makes it necessary that our party act energetically in order that it play a decisive role." The draft program perceived the "petty bourgeois" nature of the agrarian revolt and prophesied that a forthcoming

---

[22] Irving Howe and Lewis Coser, *The Communist Party: A Critical History* (New York, 1962) makes no reference to the agrarian work of the party. Nathan Glazer, *The Social Basis of American Communism* (New York, 1961), 8-9, specifically excludes farmers from the groups he considers. The projected volume in the *Communism in American Life* series dealing with the thirties has not yet appeared.

[23] Eight articles in 1928-29 dealt with farm problems. Most comprehensive of these were ["Harrow"], "The Factory Farm: A Discussion Article on the Party and the Farm Problem," *The Communist*, VII (December, 1928), 761-69, and A. B. Richman, "The Economics of American Agriculture," *ibid.*, VIII (January-February, 1929), 88-94.

[24] *The Communist*, IX (February, 1930), 104-20; (March, 1930), 280-85; (April, 1930), 359-75.

"clamorous agrarian discontent" would involve farm owners, not agricultural wage workers. In view of this the party should "point out to the agrarian petty bourgeoisie the robber side of finance capital, its alliance with the rich farmer capitalist against the poor and middle farmers . . . its role as governmental oppressor, tax-looter and war-maker, and win the passive or active support of the poor and middle farmer for the proletarian revolution." Admittedly, the natural "petty bourgeois revolutionism" of small farmers would often result in "lamentable errors . . . but no alarm need be felt at such so long as the movement is fraternally bound in alliance with and is instructed by the revolutionary proletariat, so long as the force of its attack is thrown against capitalism." The basic organizational form for the party offensive would be farmer "committees of action," whose program would consist of "tenants' strikes, mass refusal to pay mortgages or interest upon them, taxpayers strikes and a physical struggle against foreclosure." The ultimate goal of the party was collective farming, but that would be de-emphasized for the present. The immediate task was to "bring class struggle into agriculture, uniting the agrarian poor against finance capital in actual struggle. . . ."[25] In adopting the draft program the Communist party made two vital decisions: first, to support property-owning farmers in their demands for immediate remedies and second, to attempt to guide these farmers toward revolutionary radicalism.

The mastermind of the Communist party agrarian program was Harold Ware (1890-1935). Ware's name, unfortunately, is remembered only because of the allegations of Whittaker Chambers who charged that Ware headed conspiratorial work in Washington, D.C. and organized in 1934 a secret, and what was to become a notorious, cell in the Department of Agriculture.[26] Actually Ware was an important, though little known figure who had a long career in the party. He had joined the party in 1919 as a charter member. His mother was Ella Reeve Bloor, the "grand old lady" of the American Communist party. Ware's principal

[25] *Ibid.*, 105, 117, 285, 371-72.

[26] Whittaker Chambers, *Witness* (New York, 1952), 275-443. It has sometimes been asserted that Ware was an employee of the United States Department of Agriculture in the late twenties. Dana G. Dalrymple of U.S.D.A., who has examined the files of the department, informs me that Ware's only function was as a dollar-a-year man who sent to the department from Russia occasional reports on Soviet agriculture.

interest had always been agriculture and in the early 1920's when the party showed no interest in the farmer, he had farmed in Pennsylvania. However, a letter from V. I. Lenin inquiring, "Have you no farmers in America?" prompted the party hastily to recall Ware and dispatch him on an agricultural survey of the United States.[27] Beginning in 1922 he headed a project to send American agricultural equipment and experts to the Soviet Union. He went to Russia personally and was entrusted by the Soviet government with managerial posts on four different state farms in as many seasons. While in Russia he demonstrated American farm machinery and dry farming techniques. He returned to the United States before the end of 1931 and began to tour agricultural areas of the West and Midwest. Ware was an able student of Marxist theory but preferred organizational work to membership on the party Central Committee. Personally, Ware was shy and retiring, hence uncomfortable before large audiences, but he was effective and persuasive with committees and small gatherings. His leadership was exercised behind the scenes and can be documented only from correspondence and minutes. His name was almost totally absent from party publications.[28]

The front man and field organizer for the party farm effort was Lem Harris, a close friend of Ware and an associate of his in Russia. The son of a wealthy New York family, Harris left Harvard in 1926 a dedicated "Tolstoyan idealist." He went to work on a farm and shortly afterwards journeyed to the Soviet Union, where he worked on a state farm and in an agricultural implement factory. He returned to the United States at Ware's behest and joined Ware on his 1931 tour of farming areas. Harris was a man of considerable personal charm and meticulous organizing ability; he was the most successful of any of those associated with the Communist farm program in making contacts and friendships at the grass-roots. He wrote for the party-front newspaper, organized farm conferences, and served as executive secretary of the

[27] The letter is mentioned in Ella Reeve Bloor, *We Are Many* (New York, 1940), 268. Lenin displayed a surprising interest in the problems of American agriculture. He had written in October, 1913 a treatise, "Capitalism and Agriculture in America," which was reprinted in *The Communist,* VIII (June, 1929), 313-18; (July, 1929), 395-401; (August, 1929), 473-77.

[28] Bloor, 266-79; Lem Harris, personal interview, April 11, 1962; Dana G. Dalrymple, "The American Tractor Comes to Soviet Agriculture: The Transfer of a Technology," *Technology and Culture,* V (Spring, 1964), 191-214.

Farmers' National Committee for Action, discussed below.[29]

The draft program of 1930 had predicted the coming of "clamorous farm discontent," but the party was unprepared for the sudden outbreak of a militant farmers' strike in western Iowa in August, 1932. In spite of the draft program, rural work had lagged. There was no groundwork in the strike area; the party was virtually nonexistent there. The only semblance of a party organization among farmers was the United Farmers' League, founded in 1926 as a part of the Peasants' International. The UFL could claim several thousand members in 1932, but it was regarded as largely a "red" outfit and its membership was concentrated in the radical Finnish cooperatives of northern Wisconsin, Michigan, and Minnesota. Its newspaper had lapsed and the draft program had been critical of its lack of vigor.[30]

To achieve their dual objectives of building class consciousness in rural areas and creating an agrarian wing to the proletarian movement, Ware and other leaders developed a two-part strategy. First, a nation-wide farm conference of the "rank and file" would publicize a distinctive series of Communist-conceived farmer demands. Second, leaders in the field would attempt to establish liaison with protesting farmers and their organizations in order to channel these in a genuinely radical direction.

The initial step to implement the first objective was taken on September 9, on the occasion of the mass demonstration organized in Sioux City by the Farmers' Holiday Association to coincide with the conference there of midwestern governors. "Mother" Bloor presided at a meeting of about fifty farmers from which emanated a call for a Farmers' National Relief Conference to assemble in Washington in December. Ware was present and Harris was invited to serve as executive secretary for the forthcoming conference.[31] Toward this meeting the Communist party

---

[29] Lem Harris, personal interview, April 11, 1962. *Farmers' National Weekly*, April 26, 1935. The paper was published in Chicago (January 30–November 10, 1933) and Minneapolis (January 15, 1934–August 21, 1936).

[30] Theodore Draper, *American Communism and Soviet Russia* (New York, 1960), 178-79; H[enry] Puro, "The Class Struggle in the American Countryside," *The Communist*, XII (June, 1933), 556-57. In July, 1932 party workers in Washington, D.C. began publishing a small mimeographed newspaper, *Farm News Letter*, which lasted until January, 1933.

[31] *Sioux City Tribune*, September 10, 1932; *Farm News Letter*, September 16, 1932.

directed its principal attention among farmers during the autumn of 1932.

As to the second strategic objective, the Communists established contact early in September with an independent group of Holiday farmers in Nebraska. Given the circumstances of the Holiday movement in that state, the contact proved a fruitful one. Some Nebraska farmers had joined spontaneously in the farm strike in August, but the national body of the Farmers' Holiday Association had no formal organization in the state. In both Iowa and Minnesota, where state units of the Holiday were strongest, the association had developed under the friendly aegis of the Farmers' Union, making use of that organization's publicity organs and its prestige. However, in Nebraska the Farmers' Union was dedicated solely to cooperative business activities and opposed all political involvements. Its leaders were estranged from the National Farmers' Union and the *Nebraska Union Farmer* ignored the Holiday Association.[32]

During the farm strike a few organizers from Iowa ventured across the state line in an attempt to extend their area of operations, but the initiative in Nebraska had been lost by Reno's FHA. A remarkable tiny group of farmers from the hamlet of Newman Grove, Madison County, Nebraska, operating independently of the national organization, formed their own Farmers' Holiday Association. In terms of their economic demands, the Madison County group differed little from Reno's NFHA, but the leaders believed the plan for withholding produce from market, the principal tactic of the Reno organization, could lead only to violence and failure. The principal problem, to their way of thinking, was saving delinquent farmers' property from foreclosure and they set about to organize farmers in Madison and adjoining counties for that purpose. No economic or social variables segregate Madison County or the Newman Grove vicinity from other comparable areas in eastern Nebraska or western Iowa. The main crop was hogs, the main political party was Republican and while nationality origin was diverse, it was no more so than in scores of nearby counties. Farm tenancy was not common around Newman Grove; of the half-dozen most active leaders of the Madison County Plan

---

[32] Shover, "The Farm Holiday Movement in Nebraska," *Nebraska History*, 58-60; *Nebraska Union Farmer* (Omaha), September 16, 1932. The latter was published by the Nebraska Farmers' Union.

Farmers' Holiday, all were property holders. The main point of difference was that over the years a respected elderly farmer, Andrew Dahlsten, a Socialist and freethinker, had developed and nourished a tiny coterie of farm radicals. Dahlsten had been a Populist; he was a personal friend of Arthur C. Townley, founder of the North Dakota Non-Partisan League, and during World War I Newman Grove had been a center of League activities. Dahlsten, more than any other single factor, was the moving spirit behind the development of the Madison County Plan.[33]

Communist agrarian leaders joined the Madison County FHA early in its history although the original organizing impulse came from farmers, not party members. Liaison between the party and the grass-roots movement was provided by Harry Lux, a laborer from Lincoln, who had worked with Dahlsten in the Non-Partisan League a decade earlier. Lux was a colorful figure, reared in a prairie sod-house, self-educated, a war veteran, and former member of the I.W.W. He was no Communist but had known Ware for ten years, wrote comradely letters to party headquarters, and willingly served Communist purposes. Lux had attended Mother Bloor's meeting in Sioux City and he wrote to Ware about the central Nebraska situation a few days later. Ware's chance arrival one evening in Dahlsten's farmyard had obviously been engineered in advance. Dahlsten willingly accepted Ware's offer of Communist services and a few days later when a Newman Grove contingent invaded the Nebraska organizing convention of the NFHA, Ware accompanied them.[34]

The conference called by Reno's Holiday group at Fremont, Nebraska, September 16, 1932 had an unexpected result. Reno was the keynote speaker, but Madison County delegates flooding

[33] Andrew Dahlsten, "Holiday Association of Nebraska Organized Under the Madison County Plan: Origin, Purpose, Plan of Organization and Method of Procedure" (unpublished MS, n.d., copy in writer's possession); *Norfolk* (Nebraska) *Press*, September 22, 25, November 10, 1932, January 26, 1933. Mr. and Mrs. Clarence Anderson, son-in-law and daughter of Andrew Dahlsten, personal interview, March 8, 1962.

[34] Harry Lux, personal interview, March 1-2, 1962 (a duplicate of this tape-recorded interview is held by the Nebraska State Historical Society, Lincoln); Harry Lux to the writer, May 4, 1962; Harold Ware, "Preliminary Agrarian Report [1923]," and Harry Lux to Harold Ware, September 17, 1932, files of the Farmers' National Committee for Action (private collection held by Lem Harris, New York City), hereinafter referred to as FNCA.

the meeting won approval of their set of demands, including an endorsement of the forthcoming Farmers' National Relief Conference. Unknown to Reno, the demands the convention adopted had been written by Ware. A Farmers' Union official and ally of Reno, Harry C. Parmenter, was elected state chairman but vice-president was Anton O. Rosenberg from Newman Grove.[35]

Throughout the autumn and winter months tiny Newman Grove was a storm center of radical farmer activity. On October 6, seventy-five men who styled themselves the "red army" of the Nebraska FHA dragged from a garage two trucks a sales company had reclaimed from a delinquent farmer. Loaded with farmers the trucks led a caravan to the nearby village of Elgin where a widow's chattels were being sold at foreclosure. With several hundred farmers supervising proceedings, all items offered —cattle, horses, chickens—sold for 5¢ each and there were no competing bids. The total proceeds of the sale were $5.35; the mortgagor, the receiver for the Elgin State Bank, gauged the temper of the crowd and reluctantly accepted the settlement. This was the first in a series of several hundred "penny" or "Sears-Roebuck" sales that swept the cornbelt and compelled creditors to consider carefully before initiating foreclosure proceedings.[36] A few days later Madison County Plan members halted another chattel foreclosure at Petersburg and also a tax sale at Madison. Grass-roots organizers from Madison County held meetings in township schoolhouses, in farm lots, and one meeting at the Boone County fairgrounds allegedly drew 2,500. Newman Grove was graced by frequent visits of some of the top luminaries in the American Communist party: Ware, his mother, and her husband, Andrew Omholt. Functioning in an advisory capacity, Communists wrote a program for the Madison County Plan, provided propaganda for organizational meetings, and in turn used this rank and file movement to publicize the Washington conference. Rosenberg and Lux were members of the call committee and twenty-eight from Nebraska joined the hegira to

[35] Harold Ware to L. D., September 21, 1932, FNCA files; *Sioux City Journal,* September 16, 1932.

[36] Addison E. Sheldon, *Land Systems and Land Policies in Nebraska* (Lincoln, 1936), 294-99; *Norfolk* (Nebraska) *Daily News,* October 7, 1932. "Penny auction" usually referred to a real estate foreclosure, "Sears-Roebuck sale" to a chattel foreclosure.

Milo Reno, first president of the National Farmers' Holiday Association

Milo Reno in a characteristic pose

Courtesy, Des Moines Register and Tribune

Lem Harris

Harold Ware

Some Who Would Have Led the Farmers' Rebellion

Lester P. Barlow

Ella Reeve Bloor

The Iowa National Guard arrives in Lemars, Iowa, May, 1933
Courtesy, Omaha World-Herald

The farmers' blockade: near Sioux City, Iowa, August, 1932
Courtesy, Des Moines Register and Tribune

Harry Lux (center), state organizer for the Madison County Plan Farmers' Holiday Association, and grim-faced members from Deuel County, Nebraska *Courtesy,* Omaha World-Herald

Deputies armed with nightsticks patrol a highway outside Omaha, September, 1932 *Courtesy,* Omaha World-Herald

A group attending the Farmers' National Relief Conference, Washington, D.C., December 7-9, 1932

Farmers march on St. Paul, March 22, 1933

Courtesy, Minnesota Historical Society

The farmers march on the state Capitol at Lincoln, Nebraska, February, 1933 *Courtesy, Nebraska Historical Society*

First violence in the Wisconsin Dairy Strike: Kenosha County, May, 1933

Courtesy, Wide World Photos, Inc.

Farmer pickets at Sioux City, November, 1933

Courtesy, Des Moines Register and Tribune

the capital. Newman Grove became a rendezvous for all delegates traveling to the conference from the western states.[37]

The Washington conference was designed to coordinate the spade-work of organizers at the grass-roots level with the agrarian objectives of the Communist party and hopefully to lay the groundwork for a permanent relationship. Two hundred and twenty-eight delegates traveling in truck caravans from twenty-six states met from December 7 to 10, heard grim reports of dire agricultural conditions throughout the nation, and paid visits to members of Congress. Party agrarian leaders were active in organizing and directing the conference. Harris, the conference's executive secretary, made detailed arrangements and supervised proceedings. Mother Bloor visited individually with every delegate and took the floor at crucial moments in the discussion; her son Harold Ware was one of three members of the resolutions committee. Delegations were received by both Vice-President Charles Curtis and President Hoover. The conference resolutions were read by the clerk to the Senate and were read on the floor of the House by Representative Edgar Howard of Nebraska with assistance in a parliamentary maneuver by Representative Fiorello H. LaGuardia of New York.[38] The conference adopted the following resolutions:

WE DEMAND:

I. *Federal Cash Relief:*

A. To raise all rural families to a minimum health and decent standard of living, a minimum fund of $500,000,000 must immediately be appropriated for the relief of that section of the distressed farm population in need of immediate relief, regardless of race, creed or color.

II. *Federal Relief in Kind:*

A. Food products and supplies needed for relief of city unemployed should be purchased by the Federal Government directly from the farmers at a price which will insure the cost of production plus a decent standard of living.

B. The processing and transportation of these food products and other relief supplies shall be regulated by the Federal Government so

[37] *Norfolk Press,* September 22, 25, November 24, 1932; *Farm News Letter,* September 28, 1932.

[38] *Daily Worker* (New York), December 9, 19, 1932. "Proceedings of the Farmers' National Relief Conference, December 7-10, 1932," transcript in FNCA files. *Cong. Record,* 72 Cong., 2 sess., 232, 274-75 (December 9, 1932).

as to prevent profits to the Food Monopolies and Transportation companies during the period of economic crisis.

III. *Administration of Relief for Farmers:*

A. Federal cash relief and relief in kind to be administered by local committees of farmers in each township, precinct or other local unit selected by a mass meeting of all farmers needing relief.

IV. *Government Price Fixing:*

A. A price regulating body controlled by actual consumers and producers, must be immediately elected whose function shall be to reduce prices to consumers and raise prices for all farm products sold. This adjustment to be made by deduction from the swollen profits of the profiteers who stand between field and family.

V. *The Defeat of Any Legislation Based on the Theory of "Surplus" Production:*

A. While millions of our population are undernourished through loss of purchasing power, the acceptance of the surplus theory is a crime against farmers and workers.

VI. *Credit:*

A. The enactment of legislation which will provide production credit for all farm families so as to insure a basis for national consumption at normal levels. This credit is to be administered as in Section III (above).

B. The defeat of all proposals and the repeal of all legislation now in force which provides credit only for well-to-do farmers and corporations with collateral.

VII. *Debt Holiday:*

A. A Moratorium on mortgages, interest and rents for all farmers whose volume of production has until recently sustained the farm family at a decent standard of living.

B. Cancellation of mortgages, interest, Food and Seed loans and debts for supplies and furnishing for farmers whose volume of production and economic unit has always been too small to carry the debt load and support the family at a minimum health standard. (Marginal farmers, sharecroppers and others.)

VIII. *No evictions:*

A. During the national crisis Congress must declare all foreclosures, seizures of property and evictions illegal. We farmers have no collateral, but we represent the majority of the farm population. We have at last been forced to organize and present to this Congress our final demands. If our duly elected national representatives and senators fail as did the local, county and state authorities, then we pledge ourselves to protect our fellow farmers from suffering and their families from social disintegration, by our united action.[39]

---

[39] A certified copy of the resolutions is in the FNCA files. See also *Cong.*

The resolutions followed the blueprint of the earlier draft program. They were thoroughly opportunistic and centered as much as those of the Farmers' Holiday Association upon immediate alleviation of the farmers' plight. In large measure the demands paralleled those of the Farmers' Union, including even a short cut to cost of production. The most significant difference was the omission of pleas for currency inflation, a major demand of the Union and Holiday. Such demands as a half-billion-dollar relief fund to be locally administered and cancellation of debts were far-reaching, but in totality the program was more reminiscent of Ignatius Donnelly than of Karl Marx. The Communist party was not once mentioned in the convention proceedings.[40]

In order to effect these demands, the Washington conference established a permanent organization, the Farmers' National Committee for Action. Rather than being a new farm organization, this was to be a confederation of existing groups, designed on paper to coordinate and publicize activities of component units. In another sense the FNCA was the mechanism through which the Communist party would attempt to guide the grass-roots farm movement toward a more mature radicalism. Original constituents were the Farmers' Holiday Association (Madison County Plan), the largest group; the United Farmers' League; and the United Farmers' Protective Association, a budding dairy farmer group from Pennsylvania. Anton O. Rosenberg of Nebraska was elected chairman, a purely honorary position, and Harris became the executive secretary of the Farmers' National Committee for Action. The conference laid plans for an organization newspaper, the *Farmers' National Weekly*, to begin publication in February.[41]

Delegates from the conference returned to the Midwest at the moment when the depression farm crisis was nearing its crest. Foreclosure sales were mounting to an all time high and farmers at Elgin and Petersburg, Nebraska had already demonstrated how a few determined objectors could defy creditors and "high

Record, 72 Cong., 2 sess., 232, 274-75 (December 9, 1932). A slightly altered version appeared in *Farmers' National Weekly*, February 10, 1933.

[40] "Proceedings of the Farmers' National Relief Conference," FNCA files.

[41] *Daily Worker*, December 9, 10, 12, 1932; Lem Harris, personal interview, April 11, 1962.

finance." While the farm situation worsened in the winter of 1933, the ideological buttresses that sustain orderly political processes were slowly eroding in the Midwest. Communist party leaders might well assume that in the crisis that seemed impending their party and their program could play a vital role.

Five

MORE MILITANT ALL THE TIME

You would be astonished if you could attend these township and county meetings of farmers, crowded full of militant farmers—more militant all the time. It is too bad when reactionaries try to steer backward. They have to follow the mass. I have spoken to big crowds every day this week [Ella Reeve Bloor to Lem Harris, January 21, 1933. Files of the Farmers' National Committee for Action, held by Mr. Lem Harris, New York City].

The withholding of produce and the picketing of highways were not the major manifestations of depression discontent in the farm-belt; the most important and most consequential activity of the Farmers' Holiday was the movement to restrain foreclosures of farm mortgages. The antiforeclosure movement came to a climax in January and February of 1933 when at least seventy-six penny auctions or Sears-Roebuck sales took place in fifteen farm states.[1] These months also marked the crest of rural insurgency.[2] More

[1] This estimate is based upon the number of such sales reported in the *New York Times, Daily Worker, Sioux City Tribune, Sioux City Journal, Willmar Tribune, Farm Holiday News, Farmers' National Weekly,* and in the Reno papers. By this compilation there were at least 23 antiforeclosure attempts in 1932, 117 in 1933 (76 in January and February), 34 in 1934, 13 in 1935, and 9 in 1936. Of the demonstrations in January and February, 1933, 29 occurred in Iowa, 10 in Nebraska, 12 in Minnesota, 6 in Illinois, 5 in Ohio, 4 in Kansas, and one or two each in South Dakota, Michigan, Montana, Wisconsin, Idaho, Missouri, Oklahoma, and Colorado.

[2] Pitirim Sorokin, *Social and Cultural Dynamics* (abr. ed.; Boston, 1957), 573, suggests the following measure of revolutionary intensity: (1) proportional extent of social (not merely geographical) area of the disturbance, (2) proportion of population actively involved (for and against), (3) duration of the disturbance, (4) proportional intensity (amount and sharpness of violence and importance of effects).

men were involved in antiforeclosure activity than any other single manifestation of protest. Crowds of obstructionist farmers at auctions ranged from about 100 to the crowd of 2,000 that appeared at Pilger, Nebraska in January.[3] The nucleus of activity remained northwest Iowa, but unlike the farm strike, the location of demonstrations against foreclosures dot the map of the entire farmbelt. Added to the general unrest in these months was the strike of dairy farmers in Wisconsin, where the Cooperative Milk Pool, an organization independent of the Farmers' Holiday, dumped milk and blockaded highways in an attempt to force price increases.

That farmer protest should take its most militant form in the drive to resist mortgage foreclosures was a predictable outgrowth of depression conditions. Foreclosure was a countryside tragedy: ". . . the most discouraging, disheartening experiences of my legal life have occurred," wrote a small town attorney in Iowa, "when men of middle age, with families, go out of bankruptcy court with furniture, team of horses, a wagon and a little stock as all that is left from twenty-five years of work."[4] Casper Westfield of Madison, Minnesota was a seventy-five-year-old pioneer and on January 30, 1,500 Minnesota Holiday members saved for him a farm he had worked for forty-three years. Louis Larson of Niobrara, Nebraska was not so fortunate. He was a graying man of sixty, weighing at least 300 pounds, who hitchhiked across Nebraska addressing meetings of the Madison County Plan. This hulking man often burst into tears describing how he had lost the farm that he had homesteaded. Max Cichon of Sugar Creek, Wisconsin refused to vacate his foreclosed farm and a small army of deputies armed with machine guns, rifles, and tear gas attacked the farm the evening of December 5, 1932, routed the family, and bore Cichon off to jail.[5] Monthly figures on foreclosure sales are not available, but circumstantial evidence indicates that farmer resistance was most frequent at the time when farm foreclosures were more numerous than any period in the twentieth century. Nineteen thirty-three was the peak year for involuntary sales; there were almost eight for every 100 Iowa farms compared to

[3] *New York Times,* January 22, 1933, p. 10.

[4] J. Remley Glass, "Gentlemen, the Corn Belt!" *Harpers,* CLXVII (July, 1933), 206.

[5] *Washington* (D.C.) *Daily News,* December 6, 1932.

only five in 1932.[6] Along with the increment in foreclosures, the climax of farmer protest followed another unfailing barometer of discontent: the price index. The index of Iowa farm prices, forty on the 100-point scale in January and forty-two in February, was at the low point in the depression period.[7]

Like the farm strike, the antiforeclosure movement was spontaneous, unguided by any leaders or organization. The first sale, probably, was one in Wright County, Iowa in October, 1931, even before the Farmers' Holiday Association came into being.[8] Unbeknownst to the participants, the method of the penny auctions followed that of squatters' "claim clubs" fifty years before which contrived to prevent sale of land homesteaded without benefit of title. Of the initial twenty-three sales held in 1932, most were inspired by the United Farmers' League or the Madison County Plan, the groups linked to the Communist party agrarian offensive. The United Farmers' League picketed a sale in Frederick, South Dakota in late September and attempted to halt a sale in Baraga County, Michigan in November.[9] The Madison County Plan Farmers' Holiday Association followed up its success at Elgin, October 7 by halting another chattel sale at Petersburg just one week later and preventing a tax sale at Madison, Nebraska on November 1.[10] Nevertheless, there would have been a penny auction movement even had there been no Communist-front farm organizations. Participants in the sales at Elgin, at Chili, Wisconsin on November 11 and at Granite Falls, Minnesota,

---

[6] Soth, 120.

[7] *Ibid.*, 16. "During 1933 hog prices were lower relative to parity than were prices of any other major agricultural commodity, standing at 44% as against 60% for cotton, 70% for wheat, and 73% for butter. During the winter of 1932-33 they declined to the lowest level in 50 years and corn prices also reached record lows." D. A. Fitzgerald, *Corn and Hogs Under the Agricultural Adjustment Act* (Washington, 1934), 2.

[8] *Iowa Union Farmer*, October 21, 1931.

[9] *Daily Worker*, September 30, November 22, 1932. Communist Unemployed Councils in cities took the lead in resisting urban evictions. When eviction notices arrived in the Negro sections of Chicago "it was not unusual for a mother to shout to the children, 'Run quick and find the Reds!' " St. Clair Drake and Horace R. Cayton, *Black Metropolis: A Study of Negro Life in a Northern City* (rev. ed.; New York, 1962), I, 86-87.

[10] *Norfolk Daily News*, October 7, 14, 1932; *Norfolk Press*, November 10, 1932.

January 28 all believed that their penny auctions were the first.[11]

Whether in Iowa, Michigan, or Ohio, the methods of a penny auction were the same. Sometimes, as at the Cecil Kestner farm near Deshler, Ohio, the presence of a hostile crowd was indicated by a noose which dangled from a tree or haymow. Often, as at Elgin, Nebraska, no bidders but friends and neighbors were present and by arrangement with the auctioneer, only one bid was accepted. A prospective bidder, not party to the scheme, might find himself surrounded by a cluster of hefty farmers and if he spoke, might feel a heavy hand on his shoulder and someone would mutter, "Plenty high, ain't it?"[12] In the end the final sum realized might be like the $1.90 on the $800 mortgage on the Walter Crozier farm at Haskins, Ohio or the $5.35 at Elgin, Nebraska. Farmer "bidders" would pay up and the property returned to the original owner. The Supreme Court of Nebraska held that inadequacy of bids could not justify a court in refusing confirmation of a sale.[13]

The antiforeclosure movement was most intense in the old farm strike areas of northwest Iowa, bearing out Reno's admonition to one of the participants, "In every part of the United States, from California to Maine, the Sioux City territory is known . . . it is a fact that you boys have did more to put the Farmers' Holiday on the map and fix it there for the future than any other group in the United States."[14] On January 4, 1,000 Plymouth County farmers, led by C. J. Schultz, county Holiday chairman, gathered at the courthouse in Lemars when the farm of John Johnson was offered for sale. Johnson was willing to sacrifice his farm for the $33,000 mortgage if he could avoid a deficiency judgment. Mr. Herbert Martin, attorney for the mortgagor, the New York Life Insurance Company, offered the only bid, one of $30,000. When

[11] Harry Lux, personal interview, March 1-2, 1962; Jack Witt, personal interview, March 24, 1962; Harry Haugland, personal interview, April 5, 1962.

[12] The best description of a penny auction I have encountered is an unpublished and undated manuscript written by an active participant, Mr. Harry Haugland of Watson, Minnesota, who was chairman of the Tri-County Council of Defense that functioned for four years in Chippewa, Lac Qui Parle, and Yellow Medicine Counties, Minnesota. A thermofax copy in my possession will be deposited in the library of the State University of Iowa.

[13] Nelson v. Doll, 124 Neb. 523 (1933).

[14] Reno to Lawrence Caspar, January 6, 1933.

he stated he had no authority to change the amount he was rushed by the crowd and dragged down the courthouse steps. Sheriff Ralph Rippey, who raced to Martin's aid, was rudely slapped and locked in among the crowd. On threat of his life, Martin was compelled to telegraph the company for permission to bid the entire amount. The company complied. The melee over, the mob converged on the office of Common Pleas Judge C. W. Pitts, refusing to allow him to leave until he promised to sign no more foreclosure decrees. The judge contended he was powerless but agreed to write an immediate letter to Governor Clyde Herring.[15] Schultz, who was a member of the Modern Seventy-Sixers, wrote boastfully to Milo Reno,

> We are proud of Plymouth county. We still contend Plymouth county holds the record of getting things done. We are organized for action, and we are acting. . . .
>
> Who said we can't drive the fear of God into those robbers and cut throats. . . .
>
> Our defense Council sat behind closed doors all day today while the Court House was packed with farmers. We have decided a good many things. We have got up speed, nothing can stop us now. . . .
>
> There has been no thought of going to Des Moines, only general discussion, and what might happen if we don't get legislation that is fair to agriculture. Some of our local editors were called on the carpet, and told to publish the truth or have the business ruined. . . .[16]

Outside the old disturbed areas, Minnesota farmers successfully halted three sales on a single day, January 28. President Bosch of the Minnesota Holiday addressed the 2,000 farmers present at Madison, the largest sale of the day. Disorder was less in Minnesota than in Iowa, but there was a fist fight on the courthouse steps at Fairmount where farmers tried without avail to prevent a sale on February 17. A week later the sheriff of Martin County outwitted 150 protestants by holding a sale inside the courthouse while the disrupters waited outside.[17] Governor William L. Langer of North Dakota, however, crowned proceedings by ordering the state militia to *prevent* county sheriffs from conducting foreclosure sales.[18]

---

[15] *Des Moines Register,* January 5, 1933.

[16] January 7, 1933.

[17] *Willmar Tribune,* January 30, February 18, 27, 1933.

[18] *New York Times,* March 17, 1933, p. 2; *Farmers' National Weekly,* May 1, 1933.

The Farmers' Holiday Association never formally condoned the extra-legal methods farmers used in dealing with forced sales. As in the case of the August farm strike, when followers took matters into their own hands, leaders were swept along. The farm strike had been suspended in November to allow the new administration time to fulfill its promises to agriculture. Implied was the threat that should the administration fail, the weapon might be unleashed again. Milo Reno explained to an inquirer that the primary purpose of the Holiday remained to force the recognition of the farmers' right to production costs, but as a temporary measure until legislation to raise prices and refinance loans was passed, the organization intended to use every means in its power to prevent farmers from being dispossessed of home and chattels.[19] The national directors established in late January a mysterious secret auxiliary, the Loyal Order of Picketeers, to deal with foreclosure problems. It encompassed, according to Reno, the most devoted members and was established to protect them from spying and publicity. Better, it was a way of assimilating into the Farmers' Holiday some dangerous activities for which the association did not want to claim responsibility. Units of Loyal Picketeers evidently existed in Nebraska, Kansas, Missouri, Iowa, and Illinois.[20]

The most important contribution of the Holiday Association to the antiforeclosure movement was the establishing of locally elected county Councils of Defense which would bring together debtors and creditors in an attempt to work out peaceful adjustments that would avoid dispossession. Such councils functioned for more than a year in many Iowa, Minnesota, and Nebraska counties. One correspondent reported that the council had settled 100 cases in his county. Reno estimated that in 50 per cent of the cases the Councils of Defense were able to bring about a settle-

[19] Reno to Fred M. Burns, January 5, 1933.

[20] Reno to Oren L. Herbert, February 3, 1933; to Mrs. Morris Self, February 2, 1933. It is likely this organization to which Mr. Holger Sorenson of Versailles, Missouri refers in a fascinating manuscript on the Farmers' Holiday that has been circulated among historians and government officials. It has never been published since the author, insisting that Holiday members were bound by an oath of secrecy, uses fictitious names; Milo Reno, for example, is "Mike Reynolds." Mr. Sorenson wrote me, January 29, 1962: "That was a sacred trust us Holiday men had in each other and I will never break that trust."

ment satisfactory to both parties, and in a form letter he sent in answer to many inquiries about the Holiday Association, he urged the establishing of councils as a basic part of county organizations.[21] One case so adjudicated was that of a debtor from Grand Meadows, Minnesota, Arthur E. Hoover. In reporting the success in a letter to the cousin of this farmer, the President of the United States, Reno added, "It has occurred to me that since you have been unable to accomplish anything during the last four years except to bring this cousin and other farmers nearer bankruptcy, that you would gladly make a contribution to the Minnesota and National Farmers' Holiday Association. . . ."[22]

Organizers and agitators of the Communist party were drawn by the magnetic attraction of social discontent to the disturbed Missouri Valley counties. Shortly after Christmas, Mrs. Ella Reeve Bloor arrived in Sioux City. Operating on a shoestring budget, she rented a tiny apartment, hauled from her suitcase ever-present pictures of Eugene V. Debs and Walt Whitman, and soon was serving coffee, doughnuts, and a homespun version of Marxist theory to all who were interested. Within a week she reported to Lem Harris she had held nine meetings in rural communities. In one of her weekly notes she scribbled, "You would be astonished if you could attend these township and county meetings of farmers, crowded full of militant farmers—more militant all the time. It is too bad when reactionaries try to steer backward. They have to follow the mass. I have spoken to big groups every day this week."[23] Yet despite her efforts a young worker in Sioux City reported to FNCA headquarters in March, "There is nothing official about our work. We are to a large extent outsiders as yet."[24]

The situation was different in Nebraska. Vigorous organizers for the Madison County Plan reported that two to three committees of action were being formed nightly and the group boasted a membership of 25,000.[25] The major momentum, how-

[21] *Iowa Union Farmer,* February 8, 1933; Reno to John F. Case, January 21, 1933; to Clint Aydelott, January 16, 1933.

[22] January 20, 1933.

[23] Ella Reeve Bloor to Lem Harris, January 4, 21, 1933. FNCA files.

[24] March 3, 1933. FNCA files.

[25] Lement Harris, "The Spirit of Revolt," *Current History,* XXXVIII (July, 1933), 425-27; Leif Dahl, "Nebraska Farmers in Action," *New Republic,*

ever, was pointed toward a climactic march on Lincoln scheduled for February 17. A headquarters was established in a Lincoln hotel two weeks prior to the demonstration; the state Senate and Assembly agreed to convene in special session and admit all marchers into the legislature's galleries. As speakers throughout the state urged farmers on to Lincoln, one particularly belligerent county leader predicted that there were be 30,000 demonstrators and if the legislature did not respond to their demands they would tear down the new Nebraska Capitol building brick by brick.[26]

Thus far, National Farmers' Holiday Association leaders had watched with apparent disregard the progress of the Communist-front movement in Nebraska. As of December 24, Reno knew nothing of the existence of Communist rivals in the farm movement and on January 10 he reported, "Nebraska is coming fine." [27] The appearance of the *Farmers' National Weekly* in January and the widely publicized demonstration highlighted the extent to which the national had been displaced in Nebraska. Launching a belated counterattack, F. C. Crocker, the Nebraska secretary, wrote Milo Reno early in February, "This will be the battle of my lifetime this week. It is and has been Communism pitted against Reno in Nebraska." Shortly before the day of the march, Crocker circulated a mimeographed flyer to the Holiday membership:

> Communism works in mysterious ways. In the Nebraska Farm Holiday Association, COMMUNISTIC agitators advocate no membership dues for State and National Org. work . . . Communism has built up a prejudice against State and National Holiday Officers. . . . Communists have circulated their literature . . . Communists have shown their moving pictures . . . Their Agents are with us. . . . They deny their idenity [sic] . . . They are now publishing a paper for the Nebraska farmers. . . . Communistic money from the Five Year Plan of Russia is being used to Communize the World . . . Is Russian money being used in Nebraska at this time? [28]

LXXIII (January 18, 1933), 265-66. The 25,000 figure was exaggerated. In its early organizational period the Madison County Plan had no dues and membership cards were freely distributed. This estimate is based upon the number of membership cards outstanding.

[26] *Norfolk Daily News,* February 11, 13, 1933; *Nebraska State Journal* (Lincoln), February 15, 1933; *Lincoln Star,* February 15, 1933.

[27] Reno to John Bosch, January 10, 1933; to Lem Harris, December 24, 1932, FNCA files.

[28] Crocker to Reno, February [n.d.], 1933. A copy of the flyer is in the Reno Papers.

To checkmate further the extremists' demands, Crocker met on February 12 with Thomas S. Allen, acting as governor during an illness of Charles Bryan, who penciled a proclamation declaring a temporary moratorium on farm foreclosures until emergency legislation could be enacted. Although the date had passed for introducing new bills in the current legislative session, Allen arranged with the National Association leaders to have a moratorium law introduced as special legislation by the governor.[29]

With charges of Communist domination reverberating from Lincoln newspapers, several hundred Nebraska farmers gathered at the fairgrounds the night prior to the march to draft a program. To initiate proceedings, Crocker and Harry C. Parmenter, secretary and president of the Reno group, appeared and attempted to seize control of the meeting. They were rudely ejected, but the farmers who came from most Nebraska counties to seek redress for desperate economic grievances were not inclined to accept leadership of Communists or any other outsiders. Lem Harris, appointed to the resolutions committee, struggled in vain to keep a currency inflation plank off the platform. When he attempted to show a film laudatory of Soviet agriculture, an angry uproar stopped the showing. Only the timely intervention of Mother Bloor prevented passage of a motion expelling him from the meeting.[30]

Three thousand farmers marched in orderly files down the streets of Lincoln on February 17, crowding the House chambers and swarming over the Capitol steps. They listened while one of their number read the legislators their list of demands which included cutting the state salaries by half, abolishing the state militia, and the controversial (and inconsequential at a state level) demand to inflate the currency. The farmers had a respectful hearing, but the most important fruit of their efforts was the mortgage moratorium bill, already presented by the governor as emergency legislation. It became law three weeks later.[31]

Communists followed up the success at Lincoln with small

---

[29] Crocker to Reno, February 19, 1933.

[30] *Daily Worker,* February 27, 1933; *Farmers' National Weekly,* March 3, 1933; *Nebraska State Journal,* February 16, 1933. Harris wrote Ware on February 22, "Nebraska was a bare victory. . . ." FNCA files.

[31] State of Nebraska, *House Journal,* 49 sess., 34 day (February 17, 1933); Sheldon, 298.

demonstrations at Bismarck and Pierre, but marching was not the forte of Communist associated groups alone. Governor Floyd B. Olson greeted thousands of Holiday marchers who streamed into St. Paul on March 22, telling them that the Holiday was "awakening the people of the United States," and using the occasion to attack his Republican opponents in the state Senate. Three thousand farmers concluded a Holiday meeting in Des Moines by marching to the State House to demand an embargo should the farm strike begin again, an outright and complete moratorium on foreclosures, and abolition of deficiency judgments.[32]

Few movements in the history of rural insurgency have ever achieved such specific and immediate results as the farmers' drive against mortgage foreclosures in the early months of 1933. The first response to farmer pressure was that of a group of eastern life insurance companies, holders of 42 per cent of farm mortgages in Iowa.[33] The president of New York Life Insurance Company notified Governor Herring on January 30 that his company, which owned $200 million of debt in Iowa, would suspend foreclosures in the state until legislation was approved to improve the status of debtors. The company's decision, according to the president, was motivated by two factors: first, an appeal from Governor Herring for all mortgagors to exercise restraint and second, the attorney assaulted at Lemars was a representative of New York Life.[34] The same day four other companies, Aetna, Connecticut Mutual, Phoenix Mutual, and Connecticut General Life, joined in the voluntary suspension. The following day, the Prudential Life Insurance Company, largest holder of farm mortgages in the nation, announced a suspension of foreclosure suits not only in Iowa but throughout the nation.[35]

Legislative action on behalf of farmer debtors followed closely in the wake of the insurance company suspensions in most farm states. The Farmers' Holiday Association had demanded state moratoria on further foreclosures at the governors' conference in September, 1932 and since January this had been a major demand of the organization. The first action was taken by the newly elected Democratic governor and legislature in Iowa on February

---

[32] *Willmar Tribune,* March 23, 1933; *Des Moines Register,* March 14, 1933.

[33] Jones and Durand, 83-87.

[34] *New York Times,* January 31, 1933, p. 1.

[35] *Ibid.,* February 1, 1933, p. 1.

17 when Governor Herring signed a bill permitting threatened farm owners to appeal for continuance of a foreclosure action until March, 1935.[36] The Iowa law, model for moratorium laws in other farm states, permitted an owner to retain possession of delinquent realty, paying to the mortgagor in addition to required payments and interest a fair rental value for the property to be determined by the courts. The law granted the mortgagee only a time extension; the mortgagor was forbidden to take possession, but through the rental payments he had the equivalent of possession throughout the extended period.[37] The Iowa moratorium was only a partial concession to farmer demands, but it was probably as far as the state might legally go without infringing upon rights of the creditor. It applied neither to chattel foreclosures nor to existing delinquent mortgages. H. R. Gross, editor of *Iowa Union Farmer* criticized the provision for court supervision of rental and principal payments as amounting to a virtual state of receivership for a farmer. In the succeeding year, only about half of the delinquent farmers who had the opportunity availed themselves of the moratorium law.[38]

Governor Olson of Minnesota, dissatisfied with the laggardliness of the legislature, declared a one-year moratorium by executive proclamation on February 25, an action subsequently held to be an unconstitutional exercise of executive power. A Minnesota law, parallel to that of Iowa, was signed by the governor on April 18. This statute was upheld in the famous case of *Home Building and Loan Association v. Blaisdell* (1934) when the Supreme Court of the United States by a 5-4 majority held the moratorium to be a valid exercise of the police power of the state in an emergency situation and no violation of the Constitutional prescription against legislative interference with private contracts.[39]

Other states followed the lead of Iowa. The Nebraska moratorium, signed by Governor Bryan in March, even more so than that in Iowa, was a direct response to farmer pressure. By March,

---

[36] *Ibid.*, February 18, 1933, p. 2.

[37] For a detailed description of one of the moratorium laws, see Home Building and Loan Association v. Blaisdell, 290 U.S. 251 (1934).

[38] *Farm Holiday News*, February 20, 1933; *Des Moines Register*, July 15, 1934.

[39] 290 U.S. 251 (1934).

seven other states, either by executive request or legislative action, had taken some action to restrain farm foreclosures. Other legislative benefits for hard-pressed farmers were forthcoming. In Iowa, the grace period for 1932 taxes was extended three months; in Nebraska a ceiling was placed on deficiency judgments and truck license fees reduced; Minnesota cut direct property taxes and imposed a 5 per cent tax on incomes above $10,000.[40]

Moratorium laws would probably have been enacted eventually without grass-roots agitation, but penny auctions and marches to the capital hastened and placed the "urgent" label on the political responses. Beyond the legal sphere, farm insurgency brought other results less easily calculable. It can never be known how many foreclosures were avoided through the peaceful and unpublicized arbitration of local Councils of Defense nor can it be estimated how many mortgagors, witnessing the intensity of farmers' determination, exercised caution and restraint in pressing suits.

Concessions by insurance companies and the passage of moratorium laws reflected upon the Farmers' Holiday Association more than any other single farm organization, giving it an illusion of strength greater than it possessed. Milo Reno had in no sense directed the spontaneous actions of farmers protesting foreclosures, but he was nevertheless looked upon as the voice of an aroused minority of farmers who were becoming a potent political force. Actual membership totals in the Association were never revealed—certainly Reno's estimate of 90,000 in February, 1933 was a patent exaggeration.[41] A more reliable approximation can be ventured from the financial receipts of the national office in Des Moines. The annual dues of the Holiday Association were 50¢ yearly; 12½¢ of this was remitted to the national office. Reno reported that the total receipts to the national as of August 23, 1933 had been the pitiful sum of $561.80; this would constitute the dues received from 4,494 members.[42] This number is a bare minimum, probably indicating largely Iowa dues-payers, since some state units, particularly that in Minnesota, were remiss in

[40] *Sioux City Journal,* April 28, 1933; Saloutos and Hicks, 449-50.

[41] Reno to William Hirth, February 25, 1933.

[42] Reno, "For Miss Prescott."

payments to the national office and farmers often joined local units without paying.[43] Nonetheless, the Farmers' Holiday Association had few actual members, a small body for so loud a voice of protest. The strength of the organization was in an erratic "movement element" that existed for the most part outside the formal organizational structure.

Considerable administrative confusion surrounded the establishing of a newspaper for the organization, the *Farm Holiday News*, in February. Founder of the publication was A. C. Townley, already famous in the Midwest as the organizer of the Non-Partisan League of North Dakota sixteen years earlier. Townley was a stormy petrel of a farm leader. By the thirties he had a reputation as a political opportunist, but he had almost as many friends in farming regions as he had enemies. Townley may have had aspirations for leadership in the Holiday movement or he may, as Reno later suspected, have intended to use the paper as a sounding board against Floyd B. Olson whom he opposed in the Farmer-Labor gubernatorial primary in 1934.[44] Consulting only with A. W. Ricker, editor of the *Farmers' Union Herald* (St. Paul), and with only indirect liaison with the national association, Townley published the first paper on his own initiative on February 20. Ricker warned Reno, ". . . if Townley has to put up the money and the energy both to build the paper he is pretty apt to want to run the show as far as the paper is concerned." John Bosch was even more skeptical: ". . . I do not know just what his plan is. Personally I am rather dubious of accepting Mr. Townley into our organization. Although he at one time was a tremendous power in the northwest, today he does not occupy that position. . . . I still have some faith in his honesty and integrity, yet I know that in Minnesota, he would do the organization more harm than good. . . ."[45]

The early issues bore the imprint of Townley's leadership—he was currently pushing a plan for the issuing of scrip currency to provide a medium for exchange of goods and services between organized farmers and organized workers.[46] In his usual ag-

[43] *Willmar Tribune*, October 31, 1933; Reno to John Bosch, August 22, 1933.

[44] Reno to Olson, July 17, 1933.

[45] Ricker to Reno, February 22, 1933; Bosch to Reno, February 4, 1933.

[46] *Farm Holiday News*, April 19, 1933.

gressive and independent fashion, Townley toured farm regions signing up members, soliciting subscriptions, loans, and contributions indiscriminately. He failed to keep records and within a few weeks the result was a tangle of embarrassing claims against the national association and a flood of complaints to Reno.[47] The upshot was that the national was forced in May to assume $1,600 of debt incurred by Townley, take over the paper, and politely shoulder A. C. Townley out of the picture.[48] To satisfy this debt, three times the total receipts of the national association, Reno apparently used personal funds and dipped into the treasury of the Iowa Farmers' Union. Townley harbored resentments about his rejection by Reno and other Holiday leaders—twenty years later, in the 1950's, Townley charged that the Farmers' Holiday Association had been dominated by Communists.[49]

Despite numerical weakness and internal confusions the publicity from the farm strike and antiforeclosure demonstrations awakened an interest in the Farmers' Holiday Association in states outside the Iowa-Minnesota nucleus. As has been observed, state organizations were established in North and South Dakota on the eve of the farm strike in August. While Iowa farmers were picketing highways, meetings to organize state Holiday associations were being held in Shelbyville, Indiana and Bowling Green, Ohio.[50]

Most auspicious of the state units founded in the autumn of 1932 was that in Wisconsin. Five thousand attended the initial meeting at Marshfield, heard speeches by Milo Reno and other leaders, then elected Arnold Gilberts from Dunn County the state president. Wisconsin farmers were active in antiforeclosure demonstrations and when Reno returned to the state in December he spoke to audiences at Marshfield and Menomonee Falls that numbered 3,000 and 6,000, respectively.[51] However, the Farmers'

---

[47] John Bosch, personal interview, April 1, 1962; John Chalmers, personal interview, October 21, 1961; Reno to J. C. Erp, May 24, 1933; to M. L. Amos, February 28, 1933.

[48] Reno to George Armstrong, July 6, 1933.

[49] O. M. Thomason to Milo Reno, May 24, 1933. Fred Stover (editor, *U.S. Farm News,* Des Moines), personal interview, March 25, 1962.

[50] *Des Moines Tribune,* August 24, 1932; *Sioux City Journal,* September 3, 1932.

[51] *Willmar Tribune,* September 7, 1932; *Iowa Union Farmer,* December 28, 1932.

Holiday Association was not the only organization of rural protest in Wisconsin. In April, 1932 the dairy farmers of the state had been organized by Walter Singler of Appleton into the Cooperative Milk Pool, an association independent of the Farmers' Union and the Holiday. Singler reported that in the month of January alone membership had swelled from 4,800 to 6,800.[52] Singler and Reno maintained distant though friendly relationships, but there was a basic conflict in viewpoint between the two associations. The Wisconsin Cooperative Milk Pool aimed to do on a state level what Sioux City and Omaha milk producers achieved locally—to win price advances by withholding produce. Through the winter months of 1932 the organization was building toward an impending strike call. On the other hand, the withholding movements of the Holiday Association were aimed at achieving cost of production by political, not economic, means. The Association had resolved in November to refrain from withholding activity until the new administration had the opportunity to enact agricultural legislation.

When Reno spoke at Menomonee Falls he met with leaders of the Milk Pool and promised his organization would support any group that called a strike to increase prices, but he insisted such a strike ought to be nation-wide.[53] This vague promise was not satisfactory to the Milk Pool leaders. Singler complained to Reno that a statement by Wisconsin Holiday leaders that they would not support a dairy strike caused milk buyers, assured of no farmer pressure, to cut back their prices: "That little mis-statement is costing the farmers in the state of Wisconsin about $60,000 a day now." In traveling across the state, Singler reported, he found hundreds of milk producers ready to strike. "Most of these people already belong to the Holiday Association and say they joined the organization to enforce cost of production prices and they are tired of inaction and are willing to strike with the pool any day we set." [54]

The Milk Pool inaugurated a withholding movement on February 15—the first of three milk strikes in Wisconsin in 1933. Pickets appeared on highways almost immediately and for a few

[52] Walter Singler to Reno, January 24, 1933.

[53] Reno to Singler, February 25, 1933.

[54] Singler to Reno, January 24, 1933.

lively days Wisconsin witnessed some of the most forceful and violent activity the depression had yet produced in the farmbelt.[55] Dairy farmers received no support from the Holiday Association. The Wisconsin directors went on record as opposed to the strike and cooperated only by advising their members not to "scab" on the Milk Pool. Differences between the two organizations had become so open in the days preceding the strike that Reno telegraphed Singler, "Regret your statement appearing in press reflecting on leaders of Holiday movement as we propose to give you every assistance in our power." [56]

In an attempt to restore order in Wisconsin, Governor Schmedeman called together representatives of the Milk Pool, the Holiday Association, and organized labor in Milwaukee on February 20. Milo Reno did not attend, but his personal representative, G. F. Bloss of the Iowa Farmers' Union, succeeded in working out a compromise that stands as one of the major strategic victories of the Holiday Association. The more militant Milk Pool leaders agreed to a "truce" in their strike efforts and to join with the Holiday to await whatever legislative action would be forthcoming after the new administration took office on March 4.[57] Troubles in Wisconsin were not over, but for the moment the restrained force of the Milk Pool was added to the not inconsiderable pressure that was building in the farmbelt for decisive federal government action to deal with the agricultural problem.

The Farmers' Holiday Association of North Dakota began at Jamestown, July 28, 1932. A convention at Bismarck in January announced that Councils of Defense had been established in fourteen counties and endorsed a program declaring the intention to halt foreclosures, refuse to sell products at less than cost of production, and pay no debts at existing prices. They called upon the incoming national administration to pass the Frazier bill and cost of production legislation and appoint John Simpson Secretary

---

[55] Singler to Reno, February 9, 1933; A. William Hoglund, "Wisconsin Dairy Farmers on Strike," *Agricultural History*, XXXV (January, 1961), 24-34.

[56] Arnold Gilberts to Reno, February 13, 1933; Reno to Singler, February 10, 1933.

[57] Albert Fickler to Reno, February 24, 1933; G. F. Bloss to Fred Fuss, February 28, 1933 in Reno Papers.

of Agriculture.[58] More so than any other state, the North Dakota association was intermeshed with state politics. Among its strongest supporters were Non-Partisan League veterans such as Governor William L. Langer, Senator Lynn Frazier, and William Lemke, elected to the U.S. House of Representatives in 1932. The Holiday president was Usher Burdick, who used the organization as a stepping stone to a political career. When factional fights broke out in the Non-Partisan League in the autumn of 1933, the Holiday Association championed Governor William L. Langer.[59] Buttressed by this kind of political support, the Holiday Association remained virile in North Dakota for five years.

Less publicity surrounded the South Dakota Farmers' Holiday Association, last of the strong state units in the Midwest. The president was Emil Loriks, an official of the Farmers' Union and a member of the state Senate. Judging from a report on June 7, 1933 that $800 in dues had been collected, the state probably had at the time about 1,600 Holiday members.[60]

Outside the Mississippi Valley, units strong enough to hold conventions and halt sales existed in Montana, Texas, and New Mexico. The Montana Association was organized at Great Falls in September, 1932 and sales were stopped at Circle and in Rosebud County in January and February.[61] A Texas Holiday Association was formed in a convention at Waco, April 26, 1933. Chairman of the board of directors was Judge L. Gough of Amarillo and the president was George Armstrong, president of the Texas Steel Company of Fort Worth. Armstrong was the author of several books on finance and he invited the Ku Klux Klan to participate in the Waco convention.[62] A convention at Portales on January 27 created a New Mexico Holiday Association which apparently had its principal support in Roosevelt and McQuay Counties. Reno addressed a meeting at Portales on March 18; the little association had sufficient tenacity to become

---

[58] *Farm Holiday News*, February 20, 1933; James W. Dodd, "Resolutions, Programs and Policies of the North Dakota Farmers' Holiday Association, 1932-1937," *North Dakota History*, XXVIII (April-July, 1961), 107-17.

[59] C. C. Simonson to Reno, October 2, 1933.

[60] *Farm Holiday News*, June 23, 1933.

[61] Thomas Horsford to Reno, January 3, 17, March 1, 1933; *Farm Holiday News*, February 20, 1933.

[62] Reno to N. F. Chapman, March 27, 1933; Armstrong to Hiram Evans, n.d., in Reno Papers; L. Gough to Reno, April 17, 1933.

an issue in state politics in 1934 and was represented at Holiday conventions as late as 1936.[63]

Other state units were fly-by-night affairs, their organizers a sundry group of cranks and renegades. Mrs. Morris Self, a colorful farm wife from Bowling Green, Ohio, who believed herself harassed by the Ku Klux Klan, the Farm Bureau, and conspiring clergymen, wrote sprightly and detailed letters to Reno, attended several national directors' meetings, but as president of the Ohio Holiday Association did little to advance the organization. Mr. R. P. King, the state secretary, was jailed on charges of sending threatening post cards to public officials, gratuitously signing someone else's name.[64] An ephemeral Holiday Association was promoted in upstate New York by Mr. John G. Scott of Craryville, an anarchist whose mimeographed newspaper, *Mother Earth,* called for a "Day of Jubilee" on July 4, 1933 when workers would seize control of all industries. Small units were established in Columbia, Scott's home county, and in April Reno addressed a meeting at Copake Falls. Reno accompanied Scott to Dutchess County where an attempt was made to form a Holiday camp in the home county of President Roosevelt. Scott played a prominent role in the Holiday convention at Des Moines in May, but he made no headway in his attempt to integrate the Western New York Milk Producers' Association, which called a milk strike in March, into the Holiday set-up.[65] More or less active Holiday units functioned in Oklahoma, around Kankakee in eastern Illinois, and at Noblesville in southern Indiana.[66] State organizers were appointed in California, Colorado, Kansas, Maryland, Michigan, Pennsylvania, and Washington.[67] The Farmers' Holi-

[63] J. P. Bartlett to Reno, February 10, March 10, 1933; Reno to O. H. Taber, March 8, 1933; Taber to Reno, April 4, 1933. *Farm Holiday News,* November 1, 1934.

[64] Mrs. Morris Self to Reno, February 2, March 23, April 14, July 10, 1933; R. P. King to Reno, November 3, 1933.

[65] Scott to Reno, February 1, March 27, April 3, 1933; Reno to Scott, April 3, 1933. *Iowa Union Farmer,* April 5, 1933.

[66] R. L. Rickerd (Oklahoma) to Reno, January 30, 1933; Reno to Charles E. Brandow (Illinois), November 5, 1933; Don Johnson (Indiana) to Reno, February 16, 1933. *Indianapolis* (Indiana) *Star,* April 14, 1933.

[67] Reno to W. H. Green (California), February 3, 1933; to O. H. Marquis (Colorado), October 8, 1933; to H. P. Anderson (Kansas), February 7, 1933; to George D. Iversen (Maryland), March 13, 1933; to H. P. Aitkin (Michigan), March 24, 1933; to Sam Schaeffer (Pennsylvania), April 14, 1933; to Jim Lydon (Washington), April 30, 1933.

day Association made no attempt to organize in Missouri because of the presence there of the Missouri Farmers' Association, whose president, William Hirth, who had been allied with Reno in the Corn Belt Committee, was cordial if not entirely sympathetic with the Holiday movement.[68]

The apex in the strength and importance of the Farmers' Holiday Association came when 3,000 delegates assembled for a special national convention at Des Moines on March 12, 1933. Behind the Holiday Association stood the threat of a renewed farm strike, held in abeyance only until it could be determined if the new administration would carry out its pledges to agriculture. Sustaining the Holiday was the Wisconsin Cooperative Milk Pool, which had reluctantly called a strike truce to join the Holiday in waiting. Behind both of them was an army of aroused farmers who in fifteen states had halted at least eighty-one farm sales and won concessions from insurance companies and lawmakers. Whether this was a unified force not even the leaders of the Holiday Association knew. The protesting farmers represented at Des Moines had an importance beyond their numerical minority in the farm states. In a critical period of drift, with legislation stalled in a lame duck Congress, the only determined initiative to remedy immediate problems had been taken by the Farmers' Union, the Holiday Association, and independent groups of insurgent farmers. The American Farm Bureau Federation could not claim a fraction of the accomplishments of the agrarian radicals in these months of crisis.

The purpose of the Des Moines meeting was to consolidate unrelated manifestations of dissent into a political program that would be presented to the Congress supported by a united front of the most militant elements in American agriculture. The convention met but eight days after a new president had been inaugurated and a special session of Congress had convened. Consistent with the original political purposes of the Holiday Association, the resolutions adopted concluded with the declaration that if Congress had not accorded the farmers "legislative justice" by May 3, a nation-wide farm strike would be called.[69]

The manifesto adopted at Des Moines was the clearest state-

[68] Reno to Hirth, February 25, 1933; Hirth to Reno, February 28, 1933.

[69] *Des Moines Register,* March 13, 14, 1933; *New York Times,* March 19, 1933, Sec. 4, p. 7.

ment rendered of the objectives of the Farmers' Holiday Association. It underscored the role of the association as the strong-arm auxiliary of the Farmers' Union, for the resolutions affirmed point by point, but in more forceful language, the political programs adopted by the Union in November.[70] The delegates declared:

We are loyal American citizens who believe in our country and its institutions and we are proud of its history. We do not desire to seek redress of our wrongs and grievances through force except as a last resort. But we are free men and we refuse to become the serfs and slaves of the usurer and money king.

A universal bank holiday has been declared for the protection of banks. Unless we receive legislative justice by May 3, 1933, we shall then prepare for a marketing strike within 10 days. . . .

We refuse to pay interest, debts or taxes until the dollar is made to serve as an honest measure of value.

We demand a national moratorium on foreclosures of farm and city property by executive order in the same way as the bank holiday.

We demand passage of the Frazier Bill.

We demand the federal government take over the banking and currency system as a public utility.

We demand legislation to assure farmers cost of production.

We demand a steeply graduated income tax and cessation of issuance of tax exempt securities.

We demand passage of the soldiers' bonus.

We advise Congress not to go to the expense of hearings on the non-sensical domestic allotment plan.

We demand that farmers ought to be represented in the drafting of new agricultural legislation.[71]

The resolutions were presumptuous and the Holiday Association was overasserting the power it possessed, but the important thing in March, 1933 was that there was no way of testing how meaningful its threats really were. Who could know but that these demands crystallized extensive grass-roots sentiment for drastic currency inflation. Never had a national administration been subjected to such potential and unknown pressure from the farmbelt. The threats had immediate relevance, for at the very time the Holiday Association was meeting, the new Secretary of Agriculture, Henry A. Wallace, was presenting a bold new farm relief bill to the President.[72] John Simpson, president of the Farmers' Union, who testified just a week later before the Senate

[70] *Iowa Union Farmer,* March 16-30, 1933.

[71] *Farm Holiday News,* March 22, 1933.

[72] *New York Times,* March 12, 1933, p. 1.

Agriculture Committee, had every reason to stress the potential force of the threat in the cornbelt, but given the events since January his statement was no exaggeration:

Mr. Simpson. . . . that was the biggest farm convention I ever attended. There were thousands of farmers there from seventeen states. . . . The Holiday Association organized on the basis, . . . of saying they will stop the minute you pass a bill that provides for cost of production for farmers. They are to keep up their efforts until Congress gives them a bill that provides cost of production, just like the railroads get, just like the telephone, etc. gets. They are an organization that has the spirit of 1776. They have taken courts by the arm and led them into a room where there was no telephone, and locked them up, and then told the sheriff he was needed at the other end of the county, and he went, and they have thrown ropes around the necks of the attorneys that brought the foreclosure proceedings, until they begged for mercy and made settlements according to the terms of the farmers.

The Chairman (Senator E. D. Smith, D-S.C.). Are there any symptoms that they are moving toward Washington? (Laughter)

Senator Frazier. Give them time!

Mr. Simpson. . . . No, they will not march on Washington.[73]

---

[73] U.S., Congress, Senate, Committee on Agriculture and Forestry, *Hearings on H.R. 3835, Agricultural Emergency Act to Increase Farm Purchasing Power,* 73 Cong., 1 sess., 1933, 116.

Six

THE RESTORATION OF AGRICULTURE

We favor the restoration of agriculture, the nation's basic industry; better financing of farm mortgages through recognized farm bank agencies at low rates of interest on an amortization plan, giving preference to credits for the redemption of farms and homes sold under foreclosure.

Extension and development of Farm Cooperative movement and effective control of farm surpluses so that our farmers may have the full benefit of the domestic market.

The enactment of every constitutional measure that will aid the farmers to receive for their basic farm commodities prices in excess of cost [The Democratic Platform of 1932, June 30, 1932, *Proceedings* of the Democratic National Convention of 1932, 146].

John Simpson reported enthusiastically to the delegates at the Farmers' Union convention a few days following the election of 1932, "The farmers of this nation won a wonderful victory. . . . The platform of the successful party pledged to see that farmers are refinanced at lower rates of interest, and long payments on the principal. It pledged to do everything possible under the constitution to see that farmers get cost of production." [1]

Had the Democratic party and its candidate specifically pledged themselves to cost of production prices as conceived by the Farmers' Union? Simpson talked with Governor Roosevelt in Albany in April, 1932 and left the interview convinced that Roosevelt "had agreed to placing a cost of production plank in

[1] Quoted in *Willmar Tribune,* November 15, 1932.

the Democratic platform."[2] The actual platform plank calling for "the enactment of every constitutional measure that will aid the farmers to receive for their basic farm commodities prices in excess of cost"[3] was hailed by Simpson as a fulfillment of the candidate's promise. As further evidence that the platform really meant "cost of production," Simpson, who was chairman of the Oklahoma delegation at the Chicago convention, avowed that he was the author of the agriculture plank and this claim was endorsed by Senator Burton K. Wheeler.[4] Convinced that a definite promise had been made, Simpson, Reno, and William Hirth[5] of the Missouri Farmers' Association campaigned vigorously for Roosevelt in the farm states.

Roosevelt's ambiguous statements on agriculture in the campaign did not disabuse cost of production zealots of their belief. In a letter to Milo Reno, Roosevelt declared that his understanding of equality for agriculture was a price that would give the farmer a wage comparable to that received by city workers and a return on capital investment equal to that of any other investor. He added, however, that there was no simple solution to the price question because of the complication of international markets, tariffs, debts, and readjustment of acreage.[6] Roosevelt's speech at Topeka on September 14, his major pronouncement on

---

[2] Fite, "John A. Simpson: The Southwest's Militant Farm Leader," *Mississippi Valley Historical Review*, 575-76.

[3] The original draft of the platform submitted to the delegates at the Chicago convention contained the words "cost of production" rather than merely "cost" as quoted at the beginning of this chapter. *New York Times*, June 30, 1932, p. 15.

[4] *Farm Holiday News*, March 15, 1934. Senator Wheeler commented to Simpson in the course of the Agriculture Committee hearings on the Farm Relief Bill, ". . . you accused me awhile ago of writing the platform. I think you wrote that portion of it and I got it into the platform." U.S., Congress, Senate, Committee on Agriculture and Forestry, *Hearings on H.R. 3835, Agricultural Emergency Act to Increase Farm Purchasing Power*, 73 Cong., 1 sess., 1933, 115. (Hereinafter abbreviated Senate Ag. Comm. *Hearings*, H.R. 3835.)

[5] Theodore Saloutos, "William A. Hirth, Middle Western Agrarian," *Mississippi Valley Historical Review*, XXXV (September, 1951), 215-32.

[6] Roosevelt to Reno (prepared by Conservation Department), August 29, 1932, Democratic National Committee Files, 1932, for Iowa. Roosevelt Library, Hyde Park, New York. Gertrude Almy Slichter, "Franklin D. Roosevelt and the Farm Problem, 1929-1932," *Mississippi Valley Historical Review*, XLIII (September, 1956), 246-47.

agriculture in the campaign, followed in general an outline prepared by Professor M. L. Wilson, but it fell short of a specific endorsement of the domestic allotment plan which Wilson advocated. He criticized waste due to bureaucracy in the Department of Agriculture and described as a "cruel joke" a plan suggested by the Federal Farm Board that "farmers plow up every other row of cotton and shoot every tenth dairy cow." To secure a necessary adjustment of farm income, he suggested a restoration of a foreign market for agricultural surplus to be achieved through the proposed Democratic tariff policy, and beyond that, some new domestic plan that would be cooperative and voluntary, capable of refinancing itself, which would not invite European economic retaliation, and would create no new bureaucracy. Despite the assertion of one distinguished authority that Roosevelt was already sold on domestic allotment, nothing was stated in the campaign that would not apply equally well either to domestic allotment or cost of production. For cost of production enthusiasts, not likely to search out subtleties in language, there were ample grounds for belief that the incoming President of the United States was on their side.[7]

The words and actions of Farmers' Union and Holiday leaders deny that they had complete confidence their wishes were soon to be fulfilled. The farm strike had been suspended in November, to be resumed should legislation prove inadequate. Milo Reno already harbored doubts by January for he confided to his friend William Hirth his fear that Roosevelt was being guided by the Farm Bureau and the Grange.[8] The reason for Reno's misgivings was the growing support for the domestic allotment plan which had been endorsed by both of these organizations. The essence of this plan was to control farm surpluses by making benefit payments to farmers who agreed to regulate their production in accordance with a government plan. The objective was to restore farm purchasing power to a "parity" level, i.e., the relationship of income to costs that had existed in the base years, 1910-14. M. L. Wilson, the plan's author, had won support from Henry I.

---

[7] *The Public Papers and Addresses of Franklin D. Roosevelt*, comp. Samuel I. Rosenman (New York, 1938-50), I, 693-711; Arthur Schlesinger, Jr., *The Crisis of the Old Order*, 424 and *The Coming of the New Deal* (Boston, 1959), 37.

[8] January 5, 1933.

Harriman, president of the United States Chamber of Commerce, Henry Wallace, Mordecai Ezekiel, an official in the Agriculture Department, and Rexford Tugwell, a Columbia professor and advisor to Roosevelt.[9] A farm relief bill, applying domestic allotment to four commodities—wheat, cotton, tobacco, and corn—by reducing acreage 20 per cent and providing for bonus payments financed by a processing tax on the first purchaser of the commodity, was passed by the lame duck House of Representatives but died in the Senate.[10] Cost of production differed from domestic allotment in that it involved no limitation of production and gauged farm prices on immediate costs, not the parity level. "I have no faith in the allotment plan," Reno declared. "I do not feel that it is being sponsored by groups friendly to agriculture, in fact, the U.S. Chambers of Commerce and others who are for this measure have fought any corrective measures in the past. . . . I do not believe the bill has any chance of becoming a law." [11]

Reno's growing skepticism was not assuaged by Roosevelt's appointment of Henry A. Wallace as Secretary of Agriculture. Reno and Wallace had been members of the Corn Belt Committee together and they had shared platforms in Iowa speaking against Hoover in 1928.[12] Publicly Reno was friendly enough: in an editorial he pledged his support and described Wallace "as well qualified as other men who have made an earnest study of the agricultural situation." His only misgiving was that the new secretary had very likely "absorbed some of the academic idiocy" of the Iowa State Agricultural College at Ames.[13] Privately Reno was more candid. He wrote Usher Burdick that Wallace was "academic, also erratic, unstable and thoroughly bureauized in his ideas." He doubted that he "would be any great improvement over the present jackass who occupies that position." [14]

Reno had other fears that transcended legislative issues and cabinet appointments. "The year 1933," he prophesized, "will per-

[9] Schlesinger, *Coming of the New Deal,* 36-37.

[10] *New York Times,* January 13, 1933, p. 1; February 3, 1933, p. 2.

[11] Reno to Oren L. Herbert, January 11, 1933.

[12] Russell Lord, *The Wallaces of Iowa* (Boston, 1947), 182.

[13] *Iowa Union Farmer,* February 27, 1933.

[14] January 25, 1933.

haps determine if we shall have a Republic or a dictatorship."[15] Reno was a "true believer," convinced that radical inflation of the currency and government guarantees of cost of production prices and cheap credit for farmers were the only ways to save a nation close to ruin.

The new President who took office on March 4 and the special session of Congress which convened five days later faced the sober task of dealing with immediate crises and restoring badly weakened confidence in the economic system. The agriculture problem was high on the list of issues to be dealt with. The original intention was that the special session should meet only to deal with the banking crisis, but Wallace and Assistant Secretary of Agriculture Rexford Tugwell convinced the President that it might be possible to pass a farm bill in the short session. Fifty farm leaders, hastily invited to Washington on March 10, agreed that prompt action and limitation of production were necessary, but reached no accord as to how curtailment would be achieved. Neither John Simpson nor any prominent leader of the Farmers' Union was present at the meeting. Specialists of the Department of Agriculture assisted by representatives from the leaders' conference quickly prepared the draft of a bill which was presented to Congress on March 16.[16]

The bill followed the outlines of the domestic allotment proposal that had died in the preceding session. In order to re-establish farm prices at parity, the Secretary of Agriculture was empowered to enter into acreage reduction contracts with individual farmers, to lease land withdrawn from production, float price-sustaining loans on stored crops, enter into marketing agreements with processors, levy processing taxes on distributors, and, if necessary, license processors in order to enforce compliance.[17] The measure, designed largely by rural economists, was the most academic of early New Deal legislation and probably the least understood. For example, John Simpson, the bill's major opponent, did not understand it: he believed it aimed at bringing

---

[15] *Iowa Union Farmer,* February 27, 1933.

[16] *New York Times,* March 11, p. 15; March 17, 1933, p. 1; Henry A. Wallace, *New Frontiers* (New York, 1934), 164-65.

[17] Wallace, 164-65; Schlesinger, *Coming of the New Deal,* 39. The text of the final bill, *U.S. Statutes at Large,* XLVIII, 31 is in *Documents of American History,* ed. Henry Steele Commager, II, 242.

prices (not the ratio of farm prices to farm costs) to the 1914 level.[18]

President Roosevelt, sending the domestic allotment bill to Congress, described it as a "new and untrod path." Administration leaders expected, nevertheless, that it would win the same ready assent as the Banking Act and the Volstead repeal.[19] Anticipations were fulfilled in the House of Representatives where the bill was reported by the Committee on Agriculture after perfunctory hearings and passed March 22, 315-98, with a rigorous gag rule in effect that limited debate to four hours and forbade amendments. Representative Bankhead crystallized the spirit when he exclaimed, "This is part of the Democratic program under the leadership at the other end of the avenue. Let us pass it here in the House. . . ."[20]

Two obstacles blocked the smooth passage of the Farm Relief bill through the United States Senate: first, an amendment to add the cost of production formula as an alternate means the Secretary of Agriculture might employ to underpin farm prices; second, an aggregation of currency expansion amendments attached as riders to the farm credit sections of the bill. Before the battle was over the new administration had suffered its first legislative defeat and had been forced into its first compromise with Congress.

Ten roll calls were taken on measures involving some aspect of cost of production or inflationary legislation. These votes have been classified by means of the Guttman scalogram, an index that arranges individual legislator's responses on an ascending scale, beginning with the roll call on which there were the fewest votes in favor of cost of production–inflation, thus isolating the most extreme proponents, on through the roll call on which there was the largest vote for the measures. Each succeeding roll call up the ascending scale classifies individual legislators in terms of the extent of their support for the two measures.[21] Such a clas-

---

[18] Senate Ag. Comm. *Hearings,* H.R. 3835, 126.

[19] *New York Times,* March 17, 1933, p. 1; Schlesinger, *Coming of the New Deal,* 8, 11; Wallace, 162.

[20] *New York Times,* March 22, p. 1; March 23, 1933, p. 1.

[21] The clearest description of the Guttman scalogram is George M. Belknap, "A Method for Analyzing Legislative Behavior," *Midwest Journal of Political Science,* II (November, 1958), 377-402. The complete scalogram for the data presented in this chapter is reproduced in John L. Shover, "Populism in the Nineteen-Thirties: The Battle for the AAA," *Agricultural History,* XXXIX (January, 1965), 17-24.

sification reveals a hard core of support by a coalition of twenty-three senators who, in 88 per cent of possible votes, supported cost of production and inflation measures. Joining them were thirteen others who voted "pro" in 63 per cent of the roll calls. Thirty-six senators constituted a powerful legislative bloc. The coalition cut across party lines and included six Republicans, all from midwestern states, and one Farmer-Laborite. In terms of sectional alignments, twenty-five came from farming and mountain states west of the Mississippi; not one came from the New England or Mid-Atlantic states. All eight senators from the four leading wheat producing states were included: eleven of the sixteen senators from mining states of the Mountain West; five of nine from leading corn and livestock producing states. Only two senators, one from Illinois and one from California, represented states with any large urban population. Included were such strong supporters of President Roosevelt in the 1932 campaign as Senators Norris, Cutting, and Wheeler. The coalition, as the following table indicates, was bipartisan, rural, western, and southern in composition.

The supplementary measures this neo-Populist coalition sought to append to the administration farm bill were the political program of the Farmers' Union and the Holiday Association. Had radical demands from the cornbelt any influence in the United States Senate? In the case of inflation, support was remarkably diffuse and predated the agitation of the Farmers' Union and Holiday Association. Included among proponents were the National Grange, a group of businessmen organized into the "Committee for the Nation," and economists such as Irving Fisher of Yale and George Warren of Cornell. Even Henry Wallace as an Iowa farm editor had called for moderate "credit expansion on a gold base." [22] On the day four inflationary amendments were added to the farm bill, a sizeable lobby of Farmers' Union leaders, including Reno and Bosch, were in Washington talking with senators and a group of them interviewed the President. In defending currency inflation, Senate advocates capitalized more on evidence

[22] Jeannette P. Nichols, "Silver Inflation and the Senate in 1933," *The Social Studies*, XXXV (January, 1934), 13-14; *Wallace's Farmer and Iowa Homestead* (Des Moines), LVII (April 2, 1932), 3. Donald R. Murphy, who succeeded Wallace as editor, reminisced in an interview, March 15, 1962, "We were all inflationists in 1933."

Inflation—Cost of Production Vote
73rd Congress, 1st session: Senate

Party	Sectional and Party Breakdown																			
	New England		Mid-Atlantic		E.N. Central		W.N. Central			South Atlantic		E.S. Central		W.S. Central		Mountain		Pacific		TOTAL[a]
	R	D	R	D	R	D	R	D	FL	R	D	R	D	R	D	R	D	R	D	
PRO (0-1)					1	2	4	2	1		3				3		5		2	23
MOD. PRO (2-3)						1	1	2			1				1		6		1	13
MOD. (4-5)					1								1				2			4
MOD. VS. (6-8)		4		2		1	1				4		5		3	1	1	2		24
ANTI (9-10)	6	1	3		1		1	1		2	2				1	1		1		20
Total	6	5	3	2	3	4	7	5	1	2	10		6		8	2	14	3	3	84

[a] Twelve senators were not included due to absences, vacancies, or ambiguous voting records.

of farm support than upon the sympathy of any other constituency. Senator Wheeler, demanding free silver, stressed as proof of popular support the threat of an unnamed midwestern farm organization to strike unless something was done to aid the farmers.[23] Senator Frazier, in the debate on his inflationary refinance measure, also referred to the strike threat and called the Holiday Association "the most militant farm movement ever organized in this country or any other country on the face of the earth."[24] Speaker Henry T. Rainey of the House of Representatives told the Rochester Chamber of Commerce, "There must be legislation for the debtor classes. They must be able to pay what they owe and we've got to relieve them. We've got a revolution on in this country from Pennsylvania to Utah by farmers who will not allow foreclosure sales."[25] Still, silver legislation was interest group legislation and, as in 1896, the demands of farmers for inflation served as a convenient rationale for those who stood to profit both more directly and more materially.[26]

Cost of production legislation was another matter. It was uniquely a Farmers' Union demand. John Simpson was the only witness to defend it in Senate hearings and the measure was, in fact, labeled the Norris-Simpson amendment. Senator McNary testified, "It was brought into this bill by the Committee . . . because it was stoutly insisted that it be made a part of the bill by Mr. John Simpson."[27] Senator Robinson, Democratic majority leader and an opponent of the proposal, stated: "It has been said here that the West favors cost of production. I remember that one Senator asserted yesterday that he did not believe it is a practicable plan, but nevertheless he is going to support it because the people in the section from which he comes have been led to believe that this is the best method for relieving the farmers."[28] Indications are strong that there was far more rural support for this plan than for the administration's proposal. One scholar who has studied the voluminous letter-files of Secretary Wallace for the first three months of 1933 reports that incoming letters favored

[23] *New York Times,* April 11, p. 32; April 18, 1933, p. 1.

[24] *Cong. Record,* 73 Cong., 1 sess., 2155 (April 22, 1933).

[25] *New York Times,* April 20, 1933, p. 3.

[26] Nichols, *The Social Studies,* 12-18.

[27] *Cong. Record,* 73 Cong., 1 sess., 1627 (April 13, 1933).

[28] *Ibid.,* 1631.

cost of production, not domestic allotment, in a ratio of about two to one.[29] A good case can be made that the pressure of the Union and the Holiday supplemented by an effective lobbying job by John Simpson brought a favorable response on cost of production legislation in the Senate Agriculture Committee and pushed the amendment through to passage on the Senate floor.

The cost of production alternative, termed by Secretary Wallace "the most notable difference of opinion in the Congressional debate," became the principal issue as soon as the Senate Committee on Agriculture took up the farm bill.[30] Secretary Wallace and administration spokesmen were given a courteous hearing, but John Simpson, the only witness defending cost of production, consumed more time than any non-government witness.

Simpson opposed domestic allotment because, he contended, reduction of acreage would not decrease production. Farmers would cultivate more intensively and output would actually increase. To enforce reduction nothing short of "an army of bureaucrats" would be necessary. He would be satisfied, Simpson said, with the pending bill if a provision was added empowering the Department of Agriculture to set an arbitrary price for farm commodities based upon actual costs of the farming operation (with a reasonable profit added). All commodities consumed domestically would be sold at this fixed price, and surplus beyond domestic needs would be disposed of on the world market.

> The mill or elevator that the farmer takes his load of wheat to is licensed by the Secretary of Agriculture and has been told that the cost of $1.10 a bushel, we will say, and suppose that 6/7 of the crop is needed for home consumption, and suppose we raised in round numbers 700 million bushels and 600 million is needed for domestic consumption and 100 is the exportable surplus, and suppose that that licensed elevator has been told that "For every load of wheat that comes in you must settle on the basis of giving $1.10 . . . a bushel for 6/7 of the load and for 1/7 of the load they would pay the farmer whatever the world price was. If a 70 bushel load of wheat comes to your home town . . . the elevator would for 60 bushels pay $1.10 or $66, and for the ten bushels that is part of the exportable surplus they would pay the average world price, 30¢ we will say, which is $3.00.

[29] Gilbert C. Fite, personal interview, April 7, 1963; see also Gilbert C. Fite, "Farmer Opinion and the Agricultural Adjustment Act, 1933," *Mississippi Valley Historical Review*, XLVIII (March, 1962), 656-73.

[30] Wallace, 167.

Only the most rudimentary enforcement machinery would be required; if the elevator paid too low a price the farmer simply filed a complaint. Neither was Simpson unmindful of the vexing problem of farm surpluses. The cost of production scheme would handle it by passing responsibility to the individual farmer. Commodities produced in excess of domestic needs could be sold at the world price or stored on the farm in the hope that in the following year the domestic market would require all crops plus surplus or that the world market price would improve. Simpson was aware that at current depressed wage levels consumers could not pay cost of production food prices, but that could be remedied by a drastic remonetization of silver.[31]

Secretary Wallace devoted considerable time in his testimony to rebutting John Simpson. He singled out as a major weakness in the cost of production plan its lack of any production control. Any farm plan had to operate against the fact that the supply of farm products as a whole exceeded the demand for them at prices which would cover the costs of most farm units. With a guaranteed price there still remained every incentive for a farmer to strive for maximum production. The domestic allotment proposal, which aimed at using production control as a lever to bring about price adjustments was more rational than the plan of the Farmers' Union to use price adjustment as a way of bringing about production control. The Secretary was correct. If, as circumstances proved, benefit payments were ineffective to reduce production, personally stored surpluses or fears of a future market glut would have been less successful.[32]

Secretary Wallace had good reason for opposing even a discretionary cost of production clause in the farm bill. To have fixed prices on domestically consumed produce at the same time subsidy payments were being made for decreased acreage would have caused hopeless confusion. If cost of production prices were guaranteed, the domestic allotment system could not function. Neither was there any security in the fact that this was only a discretionary alternative. If political pressures from the farm bloc were strong enough to force the provision into the bill, they would be strong enough to compel the Department of Agricul-

[31] Senate Ag. Comm. *Hearings,* H.R. 3835, 110, 104-28.

[32] *Ibid.,* 129-40; Wallace, 167.

ture to implement the non-mandatory powers. The Senate committee was friendly to the Secretary but unmoved by his arguments against cost of production. The administration farm bill was reported favorably, but added to it was the Norris-Simpson amendment authorizing the Secretary to fix farm prices at the cost of production level.[33]

The acrimonious Senate debate on the amendment contrasts sharply with the aura of legislative-executive harmony that surrounds descriptions of the Hundred Days. Proponents were openly defiant of the initiative of the executive branch in agricultural matters and they seemed to view cost of production less as a discretionary alternative than as a replacement for domestic allotment. Senator Huey Long launched a scurrilous attack on the brain trust in general and Mordecai Ezekiel in particular. ". . . when we mix Ph.D's and R.F.D's we are in trouble," admonished Senator Arthur Vandenberg.[34] The opposition of Democratic senators loyal to the administration and a last-minute message from Secretary Wallace were futile. The Norris-Simpson amendment was incorporated into the farm bill by a 47-41 vote. The *New York Times* noted that "administration leaders sustained the first setback in the Roosevelt legislative program today."[35]

An administration-sponsored supplement on farm credit, added during the Senate debates, provided the opportunity to attach a series of inflation proposals as riders. The Farm Credit bill called for expansion of both short- and long-term loan facilities and scaled down interest on Federal Land Bank mortgages from 5.4 to 4.5 per cent. These new mortgages were to be financed, however, not by expanded currency but by an issue of 4 per cent federal bonds. As an alternative, Senator Elmer Thomas of Oklahoma presented an amendment calling for the issuing of sufficient greenbacks to restore the commodity index to the 1921-28 average; Senator Lynn Frazier called for a refinancing of farm mortgages at 1½ per cent through issuing fiat money; Senator Burton Wheeler asked for free and unlimited coinage of silver at sixteen-to-one; Senator Huey Long demanded an increase in silver certificates through unlimited government purchase of silver at the

[33] *New York Times,* April 4, 1933, p. 2.

[34] *Cong. Record,* 73 Cong., 1 sess., 1475, 1552-53 (April 11-12, 1933).

[35] April 14, 1933, p. 1.

going market value.[36] All these inflation proposals were mandatory, leaving the President no alternative but to implement them if they were enacted into law.

The initial roll call, that on Wheeler's amendment, revealed surprising Senate support for inflationary measures. The amendment lost 43-33, but the inflationists were jubilant even in defeat. When a similar proposal had been voted on in January it had mustered only eighteen votes.[37] The Thomas omnibus amendment, combining silver remonetization, alteration of the gold content of the dollar, and issuing of greenbacks, was soon coming to a vote. It seemed likely that inflationists could unite around it to create a Senate majority. If mandatory and uncontrolled inflation was to be avoided, the initiative clearly lay with the White House.

On April 18 the President called in Senator Thomas and that evening he announced to a distressed group of his financial advisors that the gold standard would be temporarily suspended. He produced a copy of the Thomas amendment, directed Raymond Moley to have it thoroughly amended "and then give them word to pass it." [38] The administration bill was a "controlled inflation" measure, devoid of any of the printing press panaceas the rabid inflationists advocated. All mandatory provisions were removed and the decision to implement or not rested with the Chief Executive. The President was empowered to decrease the gold content in the dollar, accept silver up to $100 million in payment of foreign debts and to expand credit by $13 billion by issuing treasury notes.[39]

The President's maneuver to moderate and channel inflationist sentiment was successful. Senator Thomas emerged as an administration champion. With the exception of the ubiquitous Frazier bill, all of the amendments were withdrawn; and Frazier's bill was defeated by a decisive 44-25 vote, the largest majority aligned against any inflationary measure. The Thomas amendment, as revised by the administration, was approved by the Senate by a conclusive 64-20 margin on April 28. Every member of the infla-

---

[36] *New York Times,* March 23, p. 1; April 15, 1933, p. 1.

[37] *Ibid.,* April 18, 1933, p. 1; Schlesinger, *Coming of the New Deal,* 42.

[38] Schlesinger, *Coming of the New Deal,* 200-201; *New York Times,* April 19, 1933, p. 1.

[39] *New York Times,* April 21, 1933, p. 1.

tion bloc supported it. Immediately thereafter the Farm Relief bill passed the Senate.[40] The President may have been sympathetic from the beginning to some measure of expansion, but his conservative financial advisors were not. Controlled and discretionary inflation was an administration compromise, dictated by the power of the monetary bloc in the United States Senate and to a lesser extent, its grass-roots supporters.

The amended farm bill now met with the approval of the administration with one exception—the discretionary cost of production amendment added in the Senate. A strict cloture rule again applied when the bill was returned to the House of Representatives. The credit and inflation supplements were speedily approved, but the House, where party discipline was more rigid and rural constituencies less generously represented, rejected the Norris-Simpson amendment. The troublesome amendment was thrown into a House-Senate conference on May 4 and after a day of discussion the conferees emerged without an agreement. Reconsidered in the House on May 9, cost of production was again defeated by a decisive 283-109 majority.[41] Either the Senate had to recede or face the possibility of a long legislative deadlock.

When the cost of production amendment came for the last time to the floor of the Senate it was the final item standing in the way of the farm bill. Senator Norris, a member of the conference, advised the Senate to drop his own amendment since there was no possibility of the House accepting it. But, Norris misgauged the determination of other cost of production advocates. Senator Wheeler (who, according to the *New York Times*, was the author of the agriculture plank in the 1932 Democratic platform) declared that to recede would mean a repudiation of a pledge made in the platform. Senator Frazier asserted there would be a nation-wide farm strike if the amendment was rejected. "The Farmers' Union," Wheeler declared, "represents more dirt farmers than all the rest of the farm organizations put together. . . . We ought to have the courage to stand up and express our own views and not take the dictation of some professor down there in the Department of Agriculture." As support

[40] *Ibid.*, April 25, p. 1; April 29, 1933, p. 1.

[41] *Ibid.*, May 4, p. 1; May 5, p. 1; May 10, 1933, p. 2.

for the amendment surged, Norris reversed himself and urged the Senate to stand fast and call another conference. However, the Senate chose to avoid further delay on the farm bill and by a vote of 48-33 laid to rest the troublesome plan to guarantee to farmers cost of production prices. This amendment created by the Farmers' Union had exhausted the full range of legislative processes before its final defeat.[42]

The New Deal administration, by compromise where it was possible and resistance where it was necessary, had won congressional approval for the "new and untrod" path in farm legislation. Inflationist pressure had been channeled, it had not been checked. There was potent fuel for political opposition in the contention the Democratic party had forfeited its pledge to farmers for cost of production prices and substituted a domestic allotment plan endorsed by the Chamber of Commerce and conservative farm organizations. The farm bill confirmed the position of the first New Deal as a moderate, even conservative program that spurned strong grass-roots and congressional demands for more radical measures. Most of the renegade political protest groups of the first New Deal administration—the Long and Coughlin movements are examples—were founded in some aspect of defeated or rejected farm legislation.

No one watched the proceedings in Congress with greater absorption than Milo Reno. A Farmers' Holiday convention at Des Moines, May 3, had reaffirmed the determination to call a general farm strike in ten days. May 13—as it turned out, one day after the signing of the farm bill—was that deadline. Cost of production, the major demand of the association, had been rejected and it was clear the administration opposed it. Yet a great step forward had been taken in the direction of inflation and easier credit for farmers. The untried production control plan might improve farm prices. In light of these facts, should a farm strike be called? Reno's decision was complicated by the flaring of unprecedented violence in the Midwest which was deemed the responsibility of the Holiday Association. In his indecision, Reno telegraphed President Roosevelt on May 11, "According to press reports you are willing to do all in your power to avert farmers' strike and resultant confusion. Will you declare moratorium on farm foreclosures and executions until fair production

[42] *Ibid.*, May 11, 1933, p. 1.

costs are conceded farmer. Answer important."[43] Neither the papers of Milo Reno nor those of Franklin D. Roosevelt indicate that a reply was sent, but the next day, simultaneous with the signing of the Farm Relief Act, President Roosevelt issued the following statement: "I urge upon mortgage creditors, therefore, until full opportunity has been given to make effective the provisions of the mortgage refinancing sections of the farm relief act, that they abstain from bringing foreclosure proceedings and making any effort to dispossess farmers who are in debt to them. I invite their cooperation with the officers of the land banks, the agents of the Farm Loan Commissioner and their farmer debtors to effect agreements which will make foreclosure unnecessary."[44]

---

[43] Reno Papers.

[44] *New York Times,* May 12, 1933, p. 1.

Seven

A TIME OF TESTING

The past year has been the testing time, testing of loyalty, courage, determination and integrity of purpose. Along with our splendid success, we have also had our disappointments with exhibitions of rare courage and intelligent action. We have also had our disappointments and made, perhaps, our share of mistakes. However that may be, we have now arrived at the crucial time through which all human institutions must sometimes pass. No period in the world's history has been so pregnant with possible achievements. You people will, in the coming year, decide as to whether we shall continue to have a representative government, that is responsive to the will of the people, or a bureaucratic government, administered by politicians in the interests of that group in society that has always exploited their fellows, that group that today controls over ninety per cent of the wealth of the nation, yet, have in no way produced or earned it. You will decide whether we will continue a Republic or become a military despotism [Milo Reno, presidential address, Second Annual Convention, Farmers' Holiday Association, Des Moines, May 3, 1933. Reno Papers].

The two months that passed while Congress and the administration labored to forge a new agricultural policy were the most trying in the history of the Farmers' Holiday Association. In March, when the new President took office, the organization was at the peak of its influence. By May, its favorable image had been marred by crude and ill-timed acts of violence. Opinion might reluctantly tolerate stopping of mortgage sales, but lynch mob actions which threatened the life of a county judge were universally condemned and brought the full force of law enforcement agencies against the perpetrators.

While Reno and Holiday leaders struggled to repair the tarnished reputation of the organization, they were faced with a second problem. With the striking of cost of production from the farm bill, an action-oriented element demanded an immediate farm strike. Milo Reno equivocated; privately he believed Henry Wallace and the architects of the Agricultural Adjustment Administration had conclusively spurned Holiday Association demands, but he refused to go so far as to endorse a strike. The President was at the peak of his popularity; the new agricultural program should be given a chance to unfold before drastic action was taken.

Milo Reno was thence in the uncomfortable position, first, of having to account for violent actions of his followers which he neither sanctioned nor approved and second, of restraining his supporters from embarking upon a reckless policy of action which he himself had encouraged.

The number of farmer demonstrations to restrain foreclosures had declined from the highs of January and February; in March there were twenty-two and in April only six throughout the nation. However the penny auctions that did occur were remarkable for the sharpness of the violence they incited. As had been the case in the earlier farm strike, disorder seemed to increase as numerical participation in the movement declined.

The Communist-supported Madison County Farmers' Holiday in Nebraska had been free of any acts of violence. On March 14, Harry Lux, the state organizer, led a group of farmers and Lincoln unemployed to Wilber, county seat of Saline County, where the farm of Joe Newman was being sold at sheriff's sale. Newman's attempt to win a stay in the proceedings on the basis of the recent Nebraska moratorium law had been denied by the county court. Deputies, suspecting that interference with the sale was planned, waited expectantly in front of the courthouse, but the demonstrators slipped to the side of the building and as the hour of the sale approached they descended the basement stairs into the sheriff's office. With the office and its entrance jammed to suffocation by the 200 protestants, the sheriff was unable to reach his phone on the wall, much less leave the office to cry the sale. Meanwhile the district judge, on the outside, quickly recruited bystanders as deputies and one of them wedged open the door far enough to discharge a tear gas gun among the sheriff and his guests. Sheriff and demonstrators alike were routed and the Holiday men, thinking this ended auctions for the day in Wilber, adjourned to the Newman farm for a picnic. Unknown to them, while they enjoyed their repast, the delayed sale went on and the farm was sold. As the group traveled back to Lincoln, they were overtaken by Saline County deputies, seventeen of them were arrested and brought back to jail—the first arrests for Holiday activity in Nebraska. Lux was charged with contempt of court and inciting to riot; bail was fixed at $2,000. The Communist press gave considerable publicity to the Lux case and the party provided funds for a defense that finally carried the case to the Supreme Court of Nebraska. Only Lux was finally convicted and his thirty-day jail sentence was suspended. A *Lincoln Star* editorial attacked Lux as the "chief among agitators" while a news article noted with alarm that several of those present at Wilber carried Communist party cards.[1]

The little flare-up in central Nebraska was nothing compared to the violence that erupted in the troubled areas north of Sioux City. Difficulties began at Primghar, county seat of O'Brien County, on the morning of April 27 when a crowd of farmers gathered to await the sheriff's sale of John Shaffer's farm. Twenty-five special deputies armed with hardwood billies stood by inside the courthouse. When Chauncey W. Pitts, one of the judges of the sixth county district which included Plymouth and O'Brien Counties, arrived, he did not like the looks of the situation and pleaded in vain with O. H. Montzheimer, attorney for the creditor, to postpone the sale. Shortly after ten o'clock, a deputy sheriff stepped out on a second floor balcony and called for bids. Immediately the shout "get him" went up from the crowd below and with a sudden impulse the farmers swarmed through the courthouse doors. Heedless of the warnings of deputies who set up a line of defense at the top of the stairs, the mob rushed up the steps and in a club- and fist-swinging free-for-all were forced to retreat. While both sides nursed their wounds and prepared for

---

[1] *Lincoln Star,* March 15, 17, 1933; *Nebraska State Journal,* March 15, 1933; *Lincoln Herald,* March 24, 1933; Lux v. Nebraska, 126 Neb. 133 (1934); Harry Lux, personal interview, March 1-2, 1962. Lux nearly won for himself a further contempt citation because at his trial he persisted in referring to County Judge Kahout of Saline County as "Judge Coyote."

another onslaught, a hasty conference was arranged between Attorney Montzheimer and the defendent, John Shaffer. Shaffer proved suddenly conciliatory and accepted his creditor's offer. The settlement was announced to the crowd, but still unsatisfied and as a last act of defiance, the throng insisted that Montzheimer, the sheriff, and all the deputies file to the front of the courthouse and kiss the flag. Having had enough of farmer mobs for the day, they complied.[2]

The exact identity of the 600 militants at Primghar cannot be determined. Judge Pitts thought the crowd was composed largely of Holiday members, but he was mistaken for the Holiday was not organized in O'Brien County. The principal farmer protest organization in the county was the Modern Seventy-Sixers of Lester Barlow and the spokesman for the mob was Simon Tjossem, the county president.[3]

Primghar was only the first episode in a day of hectic activity. While the ritual on the courthouse steps was being completed, it was announced that a "hurry-up" call had been received from the Ed Durband farm near Lemars, thirty-five miles away. Durband, a delinquent tenant, had been ordered to vacate the farm several weeks previously and since that time the premises had been guarded by Plymouth County farmers to prevent the court order from being carried out. As it turned out, all was quiet on the Plymouth County front but sometime in the late afternoon it was suggested that the group gathered there go over to Lemars to sit in on a foreclosure case being tried in the common pleas court.

On the afternoon of April 27, Charles C. Bradley, senior judge of the twenty-first Iowa district, sixty years old and a bachelor, was hearing a foreclosure case initiated by a group of banks and insurance companies to test the constitutionality of the Iowa moratorium law. The hearing was in progress when about 100 farmers entered the courtroom. Although the group was orderly, Bradley was disturbed and ordered them to take off their hats and stop smoking. "This is my courtroom," he called to them. For the farmers, some of whom were still smarting from wounds received

---

[2] Judge Chauncey W. Pitts to Governor Clyde Herring, April 29, 1933, Clyde Herring Papers, Library of the State University of Iowa, Iowa City; Korgan, "Farmers Picket the Depression," 175.

[3] *O'Brien County Bell* (Primghar, Iowa), May 3, 1933.

at Primghar a few hours earlier, this was too much. Seized by a vengeful spirit of mob anger, they surged to the front of the courtroom, some slipped bandanas over their faces, and Bradley was pulled off the bench. Slapped and shaken by the mob, the helpless judge was borne out of the courthouse and thrust into the bed of a farm truck. Carried to a crossroads a mile outside of Lemars, his trousers were removed and he was threatened with mutilation. A rope was thrown across a roadside sign and pulled tight around the judge's neck as the mob demanded that he swear to authorize no more foreclosures. "I will do the fair thing to all men to the best of my knowledge," was all the judge would promise. Perhaps the dignity and bravery of Judge Bradley, perhaps the realization it was murder they were bent upon, perhaps the pleas of R. F. Starzl, editor of the *Lemars Globe-Post* ("Even some of those who were actively handling the judge seemed strange, moving like automatons, or like self-conscious actors who knew many eyes were upon them. Their eyes sought approval . . .," he later wrote), checked the insane passion of the mob. A hubcap full of grease was dumped on the judge's head, his trousers, filled with gravel, were thrown into a ditch and the mob departed, leaving the judge besmirched and nearly unconscious in the road.[4]

This was as much a lynch mob as any pack of angry men that ever disturbed the peace of a southern town. There was a hidden factor behind the assault upon Bradley—he had long been the victim of some vicious gossip that passed in Plymouth County. Rumor had it that Bradley, an unmarried man, was living in sin in Lemars. Whether this is true is not important; the farmers who attacked him believed it was true. Had another of the district judges, like C. W. Pitts, been presiding at Lemars that day it is doubtful the attempted lynching would have occurred.[5]

The attack upon Judge Bradley made front page headlines. In Washington, Republican Senator Dickinson of Iowa observed: "I presume it is an outgrowth of the farmer holiday movement." "I am afraid it indicates that trouble is starting. . . ," declared

[4] *New York Times,* April 28, 1933, p. 1; *Sioux City Journal,* April 28, 1933; *Lemars* (Iowa) *Globe-Post,* May 1, 1933.

[5] I was informed of this gossip by W. C. Daniel (former president, Woodbury County Farmers' Holiday Association) in a personal interview, March 14, 1962 and its existence was independently confirmed by Edward L. O'Connor (attorney general of Iowa in 1933) in a personal interview, March 21, 1962.

John Simpson, "I hope that we will hurry and get legislative relief through Congress. . . ." Guy Gillette, member of the House of Representatives from the district, stated, "I sympathize fully with the farmers and realize they are driven to desperation by conditions over which they have no control." [6] Late on the night of the 27th, Sheriff Rippey of Plymouth County telegraphed Governor Clyde Herring, "Situation beyond control of civil authorities. Demand militia." [7]

Before Governor Herring could act another event aggravated the already tense situation. On the morning of April 28, the sheriff of Crawford County was conducting a sale of the chattels of J. F. Shields near Denison, seventy miles south of Lemars, when a contingent of Holiday farmers from Crawford, Monona, and Shelby Counties arrived. Expecting them, the sheriff had brought along fifty deputies and state agents. The 800 farmers formed a flying wedge and running, charged into the farmyard. The outnumbered law officials fought back with fists and clubs and one who drew a gun was dunked in the water trough. The deputies were quickly overpowered, the sale ceased, and to clinch matters the attorney for the creditor was spirited away to Denison in an automobile. The sheriff immediately added his plea to that of Sheriff Rippey by demanding state militia be dispatched to Crawford County.[8]

Iowa officials had long feared an eventuality that might require sending troops into the farm regions. Attorney General Edward L. O'Connor, anxious to avoid what he considered the mistake of Governor Turner in sending troops to Cedar County without declaring martial law, had drawn up the necessary proclamations and orders to Iowa National Guard commanders; only county names were left blank. Early on the morning of April 28, Governor Herring declared martial law in Plymouth County, describing the events of the preceding day as a "vicious and criminal conspiracy and assault upon a judge while in discharge of his official duties, endangering his life and threatening a complete break-

[6] *Sioux City Journal,* April 28, 29, 1933; *New York Times,* April 28, 1933, p. 1.

[7] Herring Papers.

[8] *Sioux City Journal,* April 29, 1933; *New York Times,* April 29, 1933, p. 1; Sheriff Hugo J. Willey to Governor Herring, April 28, 1933, Herring Papers. Anton Daugaard and A. J. Johnson (two participants in the Shields sale), personal interview, March 11, 1962.

down of all law and order." [9] Crawford County was added in the afternoon. By the following morning, 220 militiamen were encamped in Lemars and a similar contingent was on the way to Denison. By evening five had been arrested in Lemars, five in Denison, and the Shields chattels sold with bayonet wielding guardsmen supervising the proceedings. Prisoners were held in special stockades closely guarded and dragnet operations were instituted to bring in the participants and ringleaders of the three riots on the 27th and 28th.[10]

The time had come for firm and decisive restraint in Plymouth County. The presence of even a token force provided an opportunity for sober second thoughts and checked what had become a heedless rush toward anarchy and bloodshed. Majority opinion in the county gave strong support to the Governor's action. An editorial in the *Lemars Semi-Weekly Sentinel* of May 3 declared,

> What happened was not due to any sudden outburst of indignation, but was the outgrowth of months of defiance of the rights of others by a small group of men with practically no effort made by officers or citizens to stop their lawless acts. This group of men began last summer to tell the other 90 percent of the county's population when they might use the highways and when they might sell their produce. The past few months they widened their activities and extended the scope of their operations. They took over the responsibility for adjusting differences between landlord and tenants and debtors and creditors in general in this and adjacent counties. They set up a "court" where persons were ordered to appear to have their business affairs adjudicated by men who had not as a rule made a success of their own business.
>
> Some functioned fairly, others developed into a form of racket where decisions were made without consideration and enforced by intimidation. Through this second group were formed the mobs which have staged demonstrations in this and adjoining counties to show that they were the law.
>
> When the Farm Holiday movement was first organized in Plymouth county it attracted a number of representative and responsible farmers who thought they might secure through it relief from the burdens under which they were staggering. When the farm strike brought defiance of law and disregard of the rights of others and failed to bring any relief many of these representative farmers severed active

---

[9] Edward L. O'Connor, personal interview, March 21, 1962. *Sioux City Journal,* April 29, 1933.

[10] *Sioux City Journal,* April 29, 1933; *New York Times,* April 30, 1933, p. 3; State of Iowa, *Biennial Report of the Adjutant General for the Years 1933 and 1934* (Des Moines, 1934), 8-15.

connection with the organization and its leadership passed into the hands of a small group of men whose unlawful activities culminated in the disgraceful affair last Thursday.

An assistant attorney general on the scene in Lemars found business and professional people relieved by the presence of troops and he expressed the opinion that "this spirit of revolution afflicted only ten to fifteen percent of the country's population." Letters and petitions to the governor from the Lion's Club, the Chamber of Commerce, the bar association, the mayor, and the town council endorsed and urged continuation of martial law.[11]

Dispatch of troops to Crawford County was less justified. Little road picketing had taken place in the county the preceding autumn and the Shield's sale was the first penny auction there.

Under the Iowa constitution, the invoking of martial law required nothing less than an insurrection—state officials were determined to find one. Original plans were to bring charges under the Criminal Syndicalism and Conspiracy Statute of 1919 where conviction carried a ten-year prison sentence. A military court of inquiry began assembling evidence and it seemed likely that trials would be by court martial unless civil courts were specifically opened by the governor.[12] Fifty-seven were in custody by May 1, among them C. J. Schultz, president of the Plymouth County Farmers' Holiday. All arrested were subjected to extensive interrogation by state attorneys and National Guard officers.[13] Investigators showed particular interest in the activities of secret societies that seemed to be cropping up in northwest Iowa. Among those arrested were the president and secretary of the O'Brien County Modern Seventy-Sixers. Considerable publicity surrounded the uncovering of a mysterious group known as the "Sons of Picketeers." [14]

---

[11] L. M. Powers to Governor Herring, May 3, 1933; copies of the letters and petitions are also in the Herring Papers.

[12] *New York Times,* May 6, 1933, p. 6; *Sioux City Journal,* April 30, 1933. It is doubtful the military could have established jurisdiction over happenings before martial law was declared. However, the advocate-general could act as a special prosecutor before a civil court.

[13] The transcript of these interrogations cannot be located. An extensive search by the writer and by Lowell Dyson of Columbia University failed to uncover it. A few pages are with the Herring Papers.

[14] *Sioux City Journal,* May 2, 1933; *New York Times,* May 6, 1933, p. 6. The organization referred to was probably the same as the Loyal Order of Picketeers. See *supra,* ch. 5.

The alleged role of the Communist party had a prominent part in the inquiries. The day the troops arrived, Major General Park A. Finley, in command at Denison, declared the whole upheaval was because of Communist activities in northwest Iowa during the preceding year.[15] One farmer arrested at Denison, suspected of Communist affiliations, emerged bruised and bloody from his interrogation by military officers.[16] On the evening of May 1 Guardsmen raided the tiny Communist party headquarters in Sioux City. Mrs. Bloor was absent but the four persons present were arrested and two typewriters and a mimeograph machine seized. Among the literature confiscated was an inflammatory mimeographed flyer, declaring that the moratorium was a dead letter, the Governor and the President slaves of the bankers and harvester trust, and urging farmers to march on Des Moines to protest the presence of troops in northwest Iowa.[17] This single sheet probably constituted the sole Communist contribution to the immediate crisis. The projected march never materialized and neither did the evidence military officials were seeking; all those arrested were released three days later without charges.[18]

To have tried the defendants before a military tribunal raised serious legal issues. The American Civil Liberties Union wired Milo Reno their offer of services to assure a public jury trial.[19] Reno consulted with Clarence Darrow in Chicago and the nation's leading criminal lawyer, despite his poor health and his seventy-six years, agreed to come to Iowa to assist the defense if criminal syndicalism and conspiracy were the basis of the indictment.[20]

As hard as the military tried to find it, there was no plan for rebellion in northwest Iowa. Had this been an insurrection the Iowa National Guard could not have prevented it. Soldier ranks were depleted when a number took leave to attend their high school graduation exercises; one prisoner was guarded by a young

[15] *Sioux City Journal,* April 30, 1933.

[16] *Sioux City Journal,* May 1, 1933; Anton Daugaard and A. J. Johnson, personal interview, March 11, 1962.

[17] A copy of the flyer is both in the Herring Papers and the files of the FNCA.

[18] *Sioux City Journal,* May 2, 4, 1933.

[19] Copy undated in Reno Papers.

[20] *Sioux City Journal,* May 4, 8, 14, 1933; Reno to Darrow, May 4, 1933.

farm lad, his own son. As was the case with Shays' or the Whiskey rebels more than a century earlier, the alleged rebellion disintegrated when a show of force was made. In a few days a spirit of levity replaced the seriousness that first prevailed at the stockades in Lemars and Denison. The prisoners pitched horseshoes, joked with the guards, and some began to see humor in the situation. "Now that was a sight," one of the prisoners related, "a bunch of old farmers lined up to take a bath with them machine guns pointed at us."[21]

The men arrested were a cross section of Iowa farmers. The average age of the eighty-six prisoners at Lemars was 42.4, of the fifty at Denison, 42.7. The military commission insisted they were mostly tenants and farm hands, not distressed owners. In one county where a systematic check was possible, this was not true. Of the ten arrested in Monona County, eight were property holders.[22]

Criminal conspiracy charges were never pressed. Evidence collected by the military was turned over to the civil courts and on May 11 the governor revoked martial law.[23] Less stringent charges of conspiracy to hinder justice and contempt of court were filed. The prosecution in the trial of the Modern Seventy-Sixers at Primghar attempted to prove its control by Communists on the basis that Lester Barlow had once been in Moscow. Twenty of the participants in the demonstration received one-day jail sentences on contempt charges and five of the leaders were given suspended sentences of one year, charged with resisting execution of a process.[24] At Lemars three men were tried on assault charges and sentenced to thirty days, another to six months.[25] The twenty-two who came to trial at Denison all were

[21] Anton Daugaard, personal interview, March 11, 1962.

[22] *Ibid.; New York Times,* May 7, 1933, Sec. 4, p. 4.

[23] *Sioux City Journal,* May 11, 1933.

[24] *O'Brien County Bell,* June 7, 28, 1933. One interesting sidelight in the Primghar trial was the revelation that there may have been a tiny core of participants who were involved in many of the most lively of Holiday activities. Jack Hamp of Granville, Iowa, a Seventy-Sixer, gave his occupation as "professional picketeer" and declared he had picketed at James and been a leader of the mob at Council Bluffs the preceding autumn; he was involved in the riot at Primghar and the assault on Judge Bradley.

[25] *Sioux City Journal,* May 14, 1933; *Willmar Tribune,* May 27, 1933; *Lemars Globe-Post,* May 29, 1933.

fined fifty dollars and sentenced to one day's imprisonment. The presiding judge told the defendants as he passed sentence, "I don't think a single man who has stood before me is a bad man or a bad citizen who should be confined as a felon. I think you have got into a situation which has given you notoriety and decreased the value of your farms."[26]

The men who mobbed Judge Bradley may have been members of the Farmers' Holiday Association, but they operated beyond the control and sanction of the leaders. Milo Reno hastened to make a public statement condemning the incident as "deplorable, in fact, revolutionary" but adding that the farmers of the community were God-fearing, law-abiding citizens who had been driven to law violation "due to some intolerable wrong under which the people have been suffering."[27] Such declarations availed little to protect the Holiday Association from the avalanche of adverse criticism heaped upon it. Attorney General O'Connor charged that racketeering methods had been used to force farmers to join. Not 20 per cent of the members in northwest Iowa had joined of their free will, he contended. Recruiters had threatened farmers that if they did not pay their dues they might find their haystacks burning some night.[28] In the wake of the disorder, right wing extremists moved into northwest Iowa. Dr. Harry Jung, honorary general manager of the American Vigilante Intelligence Federation, told an audience at Sheldon, Iowa, "Farmers who participate in open defiance of law and order are playing into the hands of the Communist International. . . ." The Holiday Association, he declared, was a subsidiary of the Communist United Farmers' League, and wherever there were milk riots, farm strikes, and labor disorders, investigation always revealed the sinister presence of Communists in the background. "Mother Ballou [sic] had been involved," he charged, in the recent outrages in Plymouth County. He urged farmers to support the American Farm Bureau Federation.[29] A few weeks later,

[26] *Des Moines Register,* May 17, 1933; Frank D. Dileva, "Attempt to Hang Iowa Judge," *Annals of Iowa,* XXXII (July, 1954), 349-50.

[27] Quoted in Kramer, 242.

[28] *New York Times,* May 6, 1933, p. 6; *Des Moines Tribune,* May 5, 1933; *Sioux City Journal,* May 6, 1933. Newspaper accounts and interviews with several individuals hostile to the Holiday have not confirmed these charges.

[29] *O'Brien County Bell,* May 31, 1933.

G. Simon Carter of the Vigilante Federation linked Milo Reno, Wallace Short, Lester Barlow, international Jews, the I.W.W., Socialists, and Communists in a common conspiracy.[30] Governor Herring told an audience at the University of Iowa, "Iowa has not gone bolshevik, and the leaders in the unfortunate occurrence in Plymouth and Crawford county were not farmers." In a private letter he advised a Farmers' Union officer not to listen to "a few mercernary [sic] leaders who are trying to line their pockets at the expense of the distressed farmers." [31] In a barely printable letter, Milo Reno replied, "I am wondering just how a man, who occupies the exalted position of Governor of the State of Iowa can adopt the methods of the ordinary cheap shyster. . . ." [32] Barbed words could not eliminate an unavoidable truth: if the Farmers' Holiday Association professed to be the spokesman for protesting farmers, it had to bear, more than any other single organization, the responsibility for their acts of violence.

The convention of the Farmers' Holiday Association at Des Moines on May 3 faced the hard decision of whether to follow through upon the bold threats of March and call for a national farm strike if Congress did not act to give farmers legislative relief. Whatever decision the delegates might reach was colored by two external factors: as of May 3 the cost of production amendment still remained in the farm bill, and, while the convention deliberated the National Guard was camped in two Iowa counties because of activities for which the Holiday was held responsible.

Milo Reno watched the progress of the agricultural bill with increasing disappointment. He could detect no evidence, he reported, that the administration intended to redeem its pledges to agriculture; the farm problem was being juggled as badly as during the Hoover administration. The agricultural bill, he predicted, was but another scheme to deliver the independent farmer into the hands of a "tyrannical and conceited" bureaucracy: the county agents of the Agricultural Extension Service. Speaking frankly to his friend William Hirth he declared, "I have no faith whatever

---

[30] *Ibid.*, June 28, 1933.

[31] *Sioux City Journal*, May 14, 1933; Herring to C. E. Carnahan, May 15, 1933 in Reno Papers.

[32] To Governor Herring, May 23, 1933.

in the gestures that are being made at the present time by the administration. It is simply the same old tactics to hand the people a little measure of relief to suppress rebellion, with no intention of correcting a system that is fundamentally wrong."[33] Reno was talking like a leader who intended to make good the threat of a farm strike.

The careful preparations for the Des Moines convention further indicate a seriousness of purpose. Rather than farmers alone, the meeting would display a broad united front, including laborers and urban debtors. The headlined speaker was A. F. Whitney, president of the Brotherhood of Railway Trainmen, and Reno traveled to Chicago to consult with E. M. Nockels, secretary of the local Federation of Labor. An invitation was extended to Congressman Fiorello LaGuardia, active at the time in forming a home-owners protective association.[34]

Nothing before May 3 gave any premonition of the changed role of Milo Reno when he confronted the 1,500 delegates. His presidential address was a model of moderation; the Roosevelt farm program was not even mentioned. When the strike decision came up for consideration, Reno took to the floor to plead with the delegates to grant power to the National Board (the presidents from each state) to determine when and if there should be a strike. He urged that no irretractable step be taken until it was determined whether the cost of production amendment would pass.[35]

Reno's unexpected stand prompted the bitterest convention battle the Holiday Association had witnessed. The conflict between a cautious leadership and an aggressive action-oriented "movement element" burst into open debate. John Scott, anarchist president of the New York Association, headed the resolutions committee that reported in favor of a strike call on May 13. Walter Singler of the Wisconsin Cooperative Milk Pool insisted that the association make a stand for all of its demands, not just the "watered-down" Norris-Simpson amendment. Others urged a

[33] Reno to George Armstrong, April 7, 1933; to William Hirth, April 23, 1933; *Iowa Union Farmer,* April 18, 1933.

[34] Whitney to Reno, April 10, 1933; LaGuardia to Reno, April 28, 1933; Reno to LaGuardia, April 30, 1933.

[35] A copy of the address is in the Reno Papers. *New York Times,* May 5, 1933, p. 11.

strike, not for legislative purposes, but to force immediate price increases. When the votes were taken the Farmers' Holiday Association reaffirmed its decision to call a nation-wide farm strike in just ten days.[36]

Milo Reno was never more ambivalent than during the tense period between May 4 and 13. On the one hand he telegraphed Senator E. D. Smith, ". . . whether strike will be called depends upon keeping the cost of production clause in farm relief bill. Farmers at high tension. They are willing to give administration chance but if cost of production is stricken hell will break loose in middle west."[37] On the other hand, just the next day he told the executive committee they were pressing for an impossible goal and reminded them that the constitution of the Holiday Association permitted them to reconsider the mandate of the convention.[38] When cost of production was eliminated from the farm bill on May 9, Reno proclaimed the strike would go on and invited every American farmer to participate.[39] Two days later he repeatedly tried to reach Governor Floyd B. Olson to solicit his aid in calling off the strike.[40]

Unaware that there was indecision in the Holiday Association, newspapers and public officials alike expressed concern over the imminent strike. Governor Schmedeman of Wisconsin informed county sheriffs they would be responsible for maintaining order; Governor Murray of Oklahoma declared everyone would be protected in their right both to buy and sell. The *New York Times* minimized dangers, estimating that only 1 per cent of farmers would participate.[41] Ed O'Neal, president of the Farm Bureau Federation, surveyed that organization's offices in eleven states

---

[36] According to one account, the convention agreed to a resolution allowing the executive committee to suspend the strike for ten days and according to another, the resolution required concurrence of the executive committee before a strike could begin. *Sioux City Journal*, May 5, 1933; *Des Moines Register*, May 5, 1933. Authority for my statement that there was a firm decision to call a strike is Reno to G. D. Woodward, May 11, 1933, in which Reno said his resolution for director approval was defeated "and the vote was practically unanimous to strike on May 13th."

[37] May 4, 1933.

[38] *Sioux City Journal*, May 6, 1933.

[39] *Ibid.*, May 10, 1933.

[40] Vince Day to Governor Floyd B. Olson, May 11, 1933, Day Papers.

[41] *Sioux City Journal*, May 12, 1933; *New York Times*, May 5, 1933, p. 11.

and reported that Wisconsin was apathetic,[42] the Minnesota leaders looking for a new cause, the Illinois movement tiny, the Nebraska organization dominated by a few radicals, and the Holiday nonexistent in other states. Most surprising was his estimate, despite the fact he was trying to disparage the strength of the movement, that 30 per cent of Iowa farmers were active in the Holiday.[43] The most objective judgment was that of Arthur E. Holt, professor of rural sociology at the University of Chicago, who with a group of students had made a 2,000-mile tour of the midwestern farm states a month earlier. "Everywhere," he reported, "we found farmers taking definite steps toward a united action in protest against existing conditions. There is developing a class consciousness among the farmers, and unless the Roosevelt farm relief measures become immediately effective, a general farm strike is certain to result." [44] Actually the potential strength of the impending strike was not known. It could be a futile demonstration or, with the relief measures of the New Deal not yet in effect, it could be an explosive demonstration that would require force or drastic legislative concessions to repel it. If past experience was a guide, the real threat was not mass support but the trouble a militant minority of participants might cause.

Through all of Milo Reno's ambivalent actions there emerges a determination to halt the farm strike at almost any cost. He appealed to President Roosevelt to use his influence to restrain farm foreclosures.[45] On the morning of May 12, with just one day to spare, Milo Reno arrived at the office of Governor Floyd B. Olson in St. Paul. The Governor had been rushed to the hospital earlier the same morning for an emergency operation, but Reno pressed the Governor's secretary, Vince Day, as to what Olson thought about the attitude of the Roosevelt administration toward

[42] A curious judgment considering the violent Wisconsin Milk Strike erupted scarcely a week later.

[43] *New York Times,* May 7, 1933, p. 32. O'Neal had been observing the Holiday movement carefully for some months. The files of the American Farm Bureau Federation contain a sizeable collection of clippings and letters from Extension agents reporting on the association's activities. Christiana McFadyen Campbell, *The Farm Bureau and the New Deal* (Urbana, 1962), 44.

[44] *New York Times,* April 12, 1933, p. 2.

[45] Reno to Roosevelt, May 11, 1933.

agriculture. Day reported that Reno "appeared anxious to find some message of hope from Washington that would warrant him calling off the strike." He informed Reno that the Governor had a very favorable opinion of the President's efforts and believed if the present experimental program was not successful, the President would test something else. In a confidential memorandum to the Governor, Day noted that Reno was undoubtedly opposed to the strike and the Minnesota and Wisconsin leaders in favor; he advised Olson that the slightest indication of his feelings would be conclusive with the Holiday officers. The Minnesota Governor, from the hospital, drafted a letter to Reno late the same day reaffirming Day's declarations of his faith in the President. A strike called now, Olson advised, without proper organization and in the face of the legislative progress already made, would create unfavorable sentiment toward relief of farmers.[46]

Upon receiving Olson's letter, Reno dispatched a telegram to all state Holiday leaders: "In view of Roosevelt's farm statement, at Governor Olson's request and leaders' advice, the executive committee declared a truce and suspension of the national holiday called for Saturday May thirteenth to a later date to be determined on effectiveness of federal legislation."[47] Reno conveyed word of the cancellation to Minnesota and Wisconsin supporters in a radio broadcast from St. Paul. In calling off the strike, Reno acted in violation of the mandate of the convention ten days earlier; even if the national board had the power to suspend, the group at St. Paul represented only a few members of that body. Reno accepted full responsibility and offered to resign the presidency of the Farmers' Holiday Association.

There was bitter opposition in the Holiday Association to the cancellation of the strike. Major state leaders like Walter Singler of the Wisconsin Milk Pool and John Bosch of the Minnesota Holiday opposed. John Chalmers, the Iowa president, disapproved the cancellation but supported Milo Reno.[48] Critical letters flooded Reno's office. A Wisconsin correspondent wrote, "The

---

[46] Vince Day to Governor Olson, May 10, 12, 1933, Day Papers; Mayer, 150-51.

[47] Reno to J. P. Bartlett *et al,* May 12, 1933.

[48] Day to Olson, May 12, 1933, Day Papers; John Chalmers, personal interview, October 21, 1961.

farmers are pretty well disgusted and I doubt if we ever have them lined up like that again. It will probably be a sort of gorilla war now, every man fighting for his home until he is either killed or captured." [49] Most embarrassing was the situation in Wisconsin where the Milk Pool called a strike despite Reno's order. Farmers and deputies battled in Waukesha County as loads of milk were dumped and tear gas released. For a week there was roadside anarchy in Walworth, Shawano, and Outagamie Counties and a farmer was killed at Saukville. Half of the Wisconsin National Guard had been called up before Walter Singler agreed to a truce on May 21.[50] This shattered the liaison between the Milk Pool and the Holiday. Singler believed that Simpson had forced Reno to withdraw because he hoped to get a "hand-out" from Washington.[51] For his part, Milo Reno made the innocuous statement, "I doubt very much if the strike was any benefit to the Milk Producers of Wisconsin. However, I wished them well and they, at least, put up a good fight." [52]

Why Milo Reno restrained his Farmers' Holiday Association from its most ambitious endeavor is, unfortunately, largely a matter of conjecture. Olson's message, the reason usually cited,[53] was obviously a contrived performance to give Reno a rationale for a step he had already decided upon. Neither was the President's public statement determinative, for Reno told reporters just hours before the cancellation that the message made no difference; he was not authorized to halt the strike.[54] Reno gave such a bewildering variety of reasons for the decision that one might suspect the real reason was never stated. Farm prices were rising in May, 1933; the index of Iowa prices rose from 58 to 68,[55] but Reno made no mention of this until after the cancellation order. Months later he wrote to a state president, who had received an earlier letter setting forth the standard reasons for the suspension,

---

[49] C. W. Busch to Reno, June 5, 1933.

[50] *Willmar Tribune*, May 18, 1933; *Sioux City Journal*, May 15, 16, 1933; Hoglund, *Agricultural History*, 24-34.

[51] Walter Singler, interview with Lem Harris, June 6, 1933, notes in FNCA files.

[52] Reno to E. N. Hammerquist, May 27, 1933.

[53] Mayer, 151; Saloutos and Hicks, 451.

[54] *New York Times*, May 13, 1933, p. 3.

[55] Soth, 16, 22.

"It would be interesting if you could know the particular thing that decided me to use my efforts to postpone the strike." [56]

Perhaps the real reason, which Reno could not state, was his fear of the militant elements among his followers such as those responsible for the outrages in northwest Iowa. His support for the strike waned noticeably after April 27. Moreover, he may have been influenced by other individuals besides Governor Olson who momentarily convinced him he was unduly pessimistic about the future of agriculture under the Roosevelt administration.[57] Whatever his motives, Reno's decision was wise. Wisconsin and Minnesota would have given some support to a farm strike, but Iowa's participation would have been governed by the presence of militia in the most loyal area. Improving prices would have deterred many sympathetic farmers from participating, thus increasing the chances of domination of the activities by reckless elements. With the memory of Lemars fresh, law enforcement officials would have tolerated little disorder on the highways. Any violence so shortly after the attempted lynching of a judge would have incurred public hostility. Finally, the gesture would have been politically ineffective. Cost of production had received fullest possible consideration and significant measures for farm relief had just won Congressional approval. The federal government could not have retreated because of threats and mob actions by a troublesome minority of the nation's farmers.[58]

---

[56] Reno to Thomas Horsford, September 10, 1933. Reno had spelled out reasons for suspending the strike in a letter to state presidents and secretaries of the National Farmers' Holiday Association, May 13, 1933 and in *Farm Holiday News*, May, 1933.

[57] One such individual was Clarence Darrow with whom Reno talked in early May. "Clarence Darrow," he later wrote, "who has had many years acquaintance with President Roosevelt is firmly convinced that the president in his own time and way, will solve our problems along socialistic lines, so we cannot wonder that the average American feels that the President should have ample time and opportunity to make good." Reno to John Scott, June 25, 1933.

[58] For a contrary view, see Kramer, 242.

Eight

THE CLASS STRUGGLE IN THE AMERICAN COUNTRYSIDE

Our party must give this revolutionary political education to the toiling farmers in the course of struggles. . . . We have to educate the American farmer toward the American "October Revolution" [H[enry] Puro, "The Class Struggle in the American Countryside," *The Communist,* XII (June, 1933), 555].

The Communist farmers' movement passed high tide in the early months of 1933. As the Farmers' Holiday Association hesitated in the face of the new federal agricultural program, so the Farmers' National Committee for Action was forced to reorient its policy toward farm rebellion.

The march on Lincoln February 17 was the high point of party activities in the farmbelt. Party leaders, enthusiastic over their achievements, were eager to push on in the wake of their successes. Few farmer leaders shared their enthusiasm. There had been personality clashes at Newman Grove, memories of the recrimination at the fairgrounds conference the night preceding the march hung heavy, and most important, the "red" charges levied by the rival national Holiday Association bore considerable weight in rural Madison County. The disposition of the rural leaders, not so hardened as party members to such attacks, was to "lie low." Andrew Dahlsten commented to Lem Harris the day following the Lincoln march, "The red scare is something awful in this state." When Harry Lux was arrested after the futile foreclosure demonstration at Wilber, the Newman Grove leaders were so lethargic in pressing his defense that Harold Ware had to come to Nebraska himself to do the job.

In addition to the disagreements between party leaders and the local farmers, Harold Ware suspected that the Nebraska grass-roots movement planned to sell out to Reno's National Holiday Association. He was conscious of the narrow ideological margin that separated those who cooperated with the FNCA from the national organization and his fears were heightened by the maneuvering of A. C. Townley, at the time editor of the *Farm Holiday News.* Talking openly with Ware, Townley described Milo Reno as "an old honest man who doesn't know what it is all about" and revealed that he was working to oust the present officers of the Nebraska Holiday, make his old friend Anton O. Rosenberg of the Madison County Plan the new president, and thus weld the Nebraska insurgents to the national organization. Actually, this was but one of Townley's independent and disruptive maneuvers that ended his career in the Holiday only a few weeks later. For Ware, however, it was determinative. Hard won gains in Nebraska seemed on the verge of slipping into the grasp of the opposing National Holiday Association. On April 14, Ware outlined to a small group of organizers for the FNCA a drastic plan to purge the Madison County Plan of its founders and call upon the membership to elect a "young and militant leadership." Several trusted party accomplices scoured rural Nebraska to rally support but only 149 farmers appeared at a convention in Loup City early in June. A remnant of the hundreds who had marched on the capital just three months earlier, the convention dutifully chose new officers, but the estrangement of Communist representatives from the Newman Grove leaders had already closed the most auspicious chapter in Communist attempts to penetrate the farm movement. A Madison County Plan Farmers' Holiday lingered for four more years, but its followers were a handful of farmers for whom the epithet "Communist" held no fears.[1]

Failure to understand farmer sensitivity to red charges and hostility to outside leadership, overestimates of the depth of rural radicalism, lingering traces of Populist inflationary sentiment,

---

[1] Lem Harris to Harold Ware, February 26, 1933; Minutes of meeting of nine leaders of the FNCA (including Ware, Mrs. Bloor, and Harris), April 14, 1933; Ware to Harry Lux, April 27, 1933, FNCA files; *Farmers' National Weekly,* July 10, 1933; *Lincoln Herald,* July 14, 1933.

suspicions that arose from failure of communication—all served to undermine the tenuous alliance of farmers and Communists. However, basic to the difficulty Communists encountered was the dual nature of their farm program. Communists fared well enough when they met farmers on their own terms and shaped pragmatic programs based upon immediate economic needs, but as soon as they advanced to their second objective, i.e., to lead the property-owning farmers toward a more mature radicalism and push forward the face of the party, they encountered resistance.

The abusive attacks of the *Farmers' National Weekly* upon Milo Reno and the Farmers' Holiday Association further encumbered Communist attempts to maintain or increase their influence among farmers. These attacks, more than any other single aspect of the agrarian policy, were the outgrowth of sectarian policies. The so-called Third Period ideology from which world communism was gradually disentangling itself in 1932-33 maintained that since the collapse of capitalism was imminent, Communist parties throughout the world should gird themselves by forming strong, independent proletarian fronts, spurning any cooperation with reformist or socialist groups. Charges that Reno was a "social fascist" who planned to "sell out" the farm movement, that his major incentive was personal profit, or that he was in league with "rich farmers, insurance companies and banker representatives," were groundless.[2] Reno might have been an enemy of communism under any circumstances, but he was a man acutely sensitive to personal criticism and the untimely attacks assured his bitter hostility and enmity. Reno, a veteran of many pitched verbal battles, was fully capable of self-defense. On March 25 he wrote Harris, "I had come to the conclusion that you were just an ordinary shyster who tries to take advantage of any situation for publicity or otherwise. . . . I am referring you to your February 10th issue of the National Farmers Weekly, in which you, on the front page attacked me personally. I would just like to know what was your incentive and why you, whose training has been in the banker, lawyer outfit, should presume to speak for agriculture." [3] Reno began inquiries into the Farmers' National Committee for Action and in a long editorial in *Farm Holiday News* of July 24

[2] *Farmers' National Weekly,* February 10, March 3, 1933.

[3] A copy of this letter is both in the Reno Papers and the FNCA files.

he singled out the FNCA as the "most conspicuous" enemy of the Holiday Association. He described Lem Harris as the "supreme boss and dictator" (Reno probably did not know of Harold Ware), charged he was really the son of a capitalist family posing under an assumed name and his concealed purpose was to mislead and confuse farmers in order to serve the interests of his father's business. Reno, who always tended to lump all enemies into one category, declared he could see no fundamental difference "in the destructiveness of this Communist group, headed by a man like Harris, and the capitalistic group, who are determined to crush out any intelligent, constructive farm movement. . . ."

Even before Reno's counterattacks, but too late for remedy, Harold Ware, the author of the attack upon Reno, realized the error in tactics. After attending the enthusiastic Holiday convention in Des Moines on March 12-13, he penned a hasty letter to his compatriots in New York, "Don't print the next issue until you get my full report. We've got to about face. I was all wet in planning to continue attack. Reno personifies their movement to the great majority of Iowa farmers—to attack him personally is to attack them—until he slips again. . . ."[4]

Attacks upon Reno momentarily abated; the Communist press urged farmers and workers to support the forthcoming farm strike and there was mention of a "united front" effort.[5] A few party workers, but not Ware or Harris, observed the proceedings at Des Moines on May 3 and 4. When the strike was finally canceled, Communists renewed their attacks on Reno by charging he had intended the betrayal from the beginning.[6]

---

[4] Ware to "All," March 14, 1933, FNCA files.

[5] *Farmers' National Weekly,* April 17, May 19, 1933; *Daily Worker,* May 11, 1933.

[6] *Farmers' National Weekly,* May 19, 1933. Mrs. Helen Wood Birney, who claimed to have been a district organizer for the Communist party in the Northwest until her withdrawal in early 1934, testified before the House Un-American Activities Committee that she attended a Communist meeting at Des Moines in conjunction with the Holiday convention in May, 1933. She rightly noted the leadership of Harold Ware and the fact that he was associated with the national party organization, not the district. She mentioned Mother Bloor, Andrew Omholt (Mrs. Bloor's husband), and Harry Luchs [sic] as having been involved. Beyond this her testimony was a maze of fact and fiction. Her principal concern was to list individuals who attended the meeting. Among them she named Harris, who was not there (he sent a telegram from Minneapolis to the convention) and several other individuals whose names occur at no time in the Communist press or FNCA files. She

At the moment when the agrarian program faltered in the face of disagreements in Nebraska and promised federal aid to farmers, workers in the field were suddenly confronted with a new difficulty. Critics in party headquarters in New York City who had observed skeptically the whole progress of the agrarian program challenged both its tactics and strategy. The concensus reached in the draft program drawn up three years earlier obscured basic theoretical differences within the Communist party. At one extreme was a faction guided by the principles of the Peasants' International *(Krestentern)*, a branch of the Comintern with headquarters in Moscow. Fearful of the bourgeois interests of property-owning farmers, this faction dismissed all rural rebelliousness as "kulak uprisings" and argued that radical activity among the farm population had to await that time when the inevitable processes of consolidation and the squeeze of the "scissors phenomena" would reduce the mass of the farm population to the status of wage workers.[7] The draft program, with its dual emphasis on supporting the immediate economic demands of farmers and then attempting to guide them toward a revolutionary position, clearly rejected the determinism of the Peasants' International.[8] Nevertheless it left unresolved impor-

stated the principal aim of the party agrarian effort was to merge all organizations into one united front. This might be a reference to the brief cooperative flurry that preceded the May convention, but as has been noted above, the Communists were trying to build their own farm organization and prevent their membership slipping away to the Holiday. She did not mention the bitter antipathy between Milo Reno and the Communists and, without proof, she asserted that the demonstrations and milk strikes were "almost entirely" sponsored by the Communist party. U.S., Congress, House of Representatives, Committee on Un-American Activities, *Investigation of Communist Activities in the Chicago Area,* 83 Cong., 2 sess., 1954, Part II, 4231-42.

[7] This conflict is summarized in an anonymous and undated MS, "Report on the Agrarian Question," FNCA files.

[8] Earl Browder, "Report of the Political Committee to the Twelfth Central Committee Plenum . . . , November 22, 1930," *The Communist,* X (January, 1931), 17-18, asserted that disagreements over agrarian policy between the *Krestentern* and "American" groups had been resolved by the draft program in the latter's favor. Nevertheless, the issue remained alive. On the eve of the Washington conference, Harrison George, an old-time radical from I.W.W. days, strongly attacked comrades who still adhered to the *Krestentern* position. The very reformist demands of propertied farmers, he insisted, were the fulcrums for building class consciousness in the countryside. "We must lead them, or fascism will—and against us," he concluded. "Causes and Meaning of the Farmers' Strike and Our Tasks as Communists," *The Communist,* XI (October, 1932), 931.

tant questions of emphasis that troubled a more important faction, a sectarian group in New York that insisted rural America was already ripe for revolution. Should the primary emphasis be in an "opportunistic" direction, as Ware tended to conceive it—winning mass rural support through advancing the immediate grievances of bourgeoisie farmers—or should principal attention be more sectarian—toward building a genuine radicalism in rural areas and creating party units in the countryside?

Writing in *The Communist* of June, 1933, Henry Puro, a member of the Political Bureau of the Party specializing in agrarian work and principal spokesman for the sectarian wing, reviewed critically the entire experience of the party in agrarian protest work. Puro did not repudiate the action of his comrades in the field, but he had a different conception of what should have been accomplished. The principal task to his way of thinking was to "politicize" the farmers' movement more rapidly. "We have to educate the American farmer toward the American 'October Revolution,'" he urged. Beginning with the Washington conference, Puro charged "the biggest mistake was that while the conference was politically inspired and organized by our Party, the party was not put forward." Resolutions calling for price fixing, being capitalistic in nature, could bring no permanent benefits to farmers and had no part in the program. Beyond the conference, Puro censured agrarian leaders for their hesitancy in building the party. Referring to the "demoralized" Nebraska movement, he asked why no progress had been made in party recruitment among the 25,000 members of the Madison County Plan. Moreover, the *Farmers' National Weekly* had erred in failing to take a stronger stand against "scoundrels like Townley and Milo Reno." [9]

A resolution on the agrarian question adopted at an extraordinary conference of the party in July, 1933 reiterated Puro's charges. The resolution was severely critical of the extreme right wing opportunism that would ignore the farmers' movement until inevitable economic developments transformed small farmers into proletarians. At the same time it found fault with the moderate form of opportunism detected in party tactics at the Washington

[9] "The Class Struggle in the American Countryside," XII, 547-55; see also Puro, "The Tasks of Our Party in the Work Among the Farmers," Speech at the Extraordinary Conference of the Communist Party, July 7-10, 1933, *The Communist*, XII (September, 1933), 880-83.

conference and thereafter. ". . . the Communists at the conference showed that they were trying to adapt themselves to the backward farmers instead of raising them to the level of the more progressive farmer. They forget that the militant farmers do not fear the Communist Party and will follow it if the Party will energetically defend their interests. . . ."[10]

To point out that the sectarian critics were demanding increased stress on the very policies that caused the most resistance when put into practice misses the essential point. The dual policy of first winning rural support and then leading farmers to a more advanced radicalism—even to party membership—was a necessity. Committed Communists had not joined in rural protest simply to achieve price and property gains for stricken farmers; to fail to advance the party in the process meant a fruitless expenditure of time and effort. Party bureaucrats from their removed vantage point were more conscious of this problem than field workers immersed in the day-to-day task of trying to conciliate rebellious farmers and shape a program that would win their support. The theoretical debate crystallized the basic dilemma on which the Communist agrarian program floundered: the attempt to guide and educate individualistic farmers which prompted resistance, hostility, and recrimination was the program's essential complement.

The critiques of agrarian policy in the summer of 1933 signaled a shift from the more opportunistic approach of the Washington conference. Workers in the field dutifully observed the admonition to "show the face of the party." The United Farmers' League, avowedly a party-controlled organization, became the main vehicle for agrarian policy. What strength the Communist program retained in rural America after the summer of 1933 was not in any mass support but in the loyal backing of a coterie of committed radical farmers who staged occasional demonstrations, attended party-sponsored farm conferences, and harassed the vestige that remained of the national arm of the Farmers' Holiday Association.

Viewing in perspective this unique effort to form an alliance with aggrieved bourgeois farmers, it is clear that Communists

---

[10] Anon. pamphlet, "The Communist Position on the Farmers' Movement," [July, 1933], 20, FNCA files.

rode the crest of the rural rebellion. The party's successes came when the protest was at its peak. In Nebraska, where the influence of Communist-inspired action can most surely be documented, the state's moratorium law would have been delayed if party organizers and their farmer allies had not impelled the march on Lincoln. The advent of Communist leaders in farm states called forth a tiny group of village radicals and isolated dissidents who were momentarily willing, after 1933, to cast their lot with communism.

Nine

MORE OR LESS OF AN ENIGMA

There are many differences of opinion as to President Roosevelt and and his program of reconstruction, in fact, the President is more or less of an enigma to many of us, but I feel that he is entitled to the break and an opportunity to demonstrate the workability of his plan. . . . To one standing on the sidelines, it is not hard to vision either of two results; one is the establishing of a state socialism or a nationalization of all industry and enterprises. . . . Again, it is not hard to see in this program of intervention in business and industrial affairs, the building of a monster autocratic, bureaucratic system that may crush all of independence and liberty of action out of our people, setting up a bureaucratic, autocratic, dictatorial government that nothing short of bloody and destructive revolution would ever overthrow. One thing is sure—it will not require any great amount of time to demonstrate which road this program will take [Milo Reno, editorial, *Farm Holiday News*, June 19, 1933].

The cancellation of the farm strike broke the high tension that had prevailed in the farmbelt since January. Speculation resulting from anticipated inflation and government assistance to agriculture brought a gradual rise in commodity prices so that in July the Iowa farm price index reached eighty-three, the highest point in two years. "I feel that we are just commencing what will amount to a real boom . . .," Milo Reno predicted.[1] The Agricultural Adjustment Administration vigorously began operations; allotment programs for cotton and wheat were underway in a few weeks and a group of Iowa farm leaders proposed a bold

[1] Soth, 16; Korgan, "Farmers Picket the Depression," 199-200; Reno to George Armstrong, July 6, 1933.

plan to deal with a threatened glut of the hog market. While the Farmers' Holiday bided its time, the initiative was shifting from the radical farm organizations to a group of leaders and administrators more orthodox and conservative.

For three months following the May crisis Milo Reno was in doubt. The phrase "more or less of an enigma" appeared more than once in his letters and speeches.[2] He spoke warmly of many New Deal measures, particularly the retreat from the gold standard. As prices rose, Reno predicted that if Roosevelt would stand fast he would perhaps go down in history as the greatest President of the United States.[3] Reno's enthusiasm, however, did not extend to the farm program. "I am more skeptical of the success of the agricultural program than I am either of his industrial or financial," he wrote in July, "not because it is more complicated . . . but because of the distressing mistake made in the selection of the Secretary of Agriculture."[4] He expressed two specific misgivings: Why, he asked, had the agents of the Agricultural Extension Service been incorporated into the federal farm machinery under the AAA? Why had the new farm loan program been financed through interest-bearing bonds rather than by expanding the currency as permitted in Title III of the Farm Relief Act?

The first question was a timely one. The relationship of the Extension service to the AAA had already provoked controversy in the Agriculture Department. Rexford Tugwell opposed M. L. Wilson's plan that state Extension directors be made the state administrators of the AAA. Around the county agents there had grown an unofficial power structure; local farm bureaus, originally created to sustain the county agents, had been incorporated into the private American Farm Bureau Federation. Supplementing this alliance between public agents and a private body were the land grant colleges. Reno was speaking as a rival Farmers' Union official, but nonetheless, the county agents and Farm Bureau represented conservative and large scale agriculture more than they represented small family farmers. They were incorporated into the new agricultural setup more by necessity than by choice;

---

[2] *Farm Holiday News,* June 19, 1933; Reno to John G. Scott, June 25, 1933.

[3] Reno to Armstrong, July 6, 1933.

[4] *Farm Holiday News,* July 25, 1933.

there was no other existing machinery which could begin the immediate operations the agrarian crisis demanded.[5]

Neither were Reno's doubts about the farm credit system baseless. The Frazier bill had been defeated and in the minds of Holiday critics the refinancing provisions of the farm bill were inadequate. Richard Bosch, ablest student of economics in the Holiday movement, pointed out that the Federal Land Bank would authorize loans only up to 50 per cent of the total land value and 20 per cent of the value of insurable buildings. With values seriously deflated, this limited the amount of credit aid; he estimated that only 17 per cent of existing mortgages could be financed, given these restrictions.[6]

While Reno puzzled over the course the organization should take, he still denied there was any weakening in the strength of the Holiday Association. He retorted to a *New York Times* report that the organization was dying that he had never had so many speaking requests and his audiences were the largest he had encountered.[7] Actually, the Holiday membership rolls included barely 4,500 persons who had paid their 12½¢ national dues. Since state units were remiss in contributing to the national, even the $200 needed to provide the 5,000 subscribers with their monthly *Farm Holiday News* caused severe financial strains in the summer of 1933. After having dipped into his own pocket, Reno wrote gloomily that unless state units took a different attitude, the national would have to disband and forsake farmers to local efforts.[8]

As if questions of meager funds and perplexing federal programs were not enough, Reno was forced to devote a major portion of his attention to problems in his own domicile. In May, a group of "bolter" policy holders in the Farmers' Union Insurance Companies, of which Reno was president, filed a petition for bankruptcy. This was the culmination of a long battle between the business and political wings of the Union that began two

---

[5] Schlesinger, *Coming of the New Deal,* 60.

[6] *Farm Holiday News,* July 24, 1933.

[7] *New York Times,* July 2, 1933, p. 8; Reno to Margaret Marshall, July 17, 1933; *Farm Holiday News,* July 24, 1933. Reno was exaggerating. His correspondence files show far more requests for information during January and February than during the summer months.

[8] Reno to James Spurlock, June 5, 1933; to F. R. Dawes, October 8, 1933; to John Bosch, August 22, 1933. This will not necessarily sustain a conclusion that the strength of the organization was declining. Small membership and inadequate finances were matters of long standing.

years earlier when opponents had wrested the Sioux City Livestock Commission house away from the Iowa Union and established a rival organization, the Progressive Farmers' Union of Iowa. Reno considered the attack an attempt to discredit him, but the bolters' contention that the company's investments were unsound and money had been squandered in political activities were not groundless. The company's securities were almost entirely in farm mortgages—a shaky financial foundation in the thirties—and company administrative funds had been tapped, although not indiscriminately, for Farmers' Holiday activities.[9] The consequence was that the Property Insurance Company went into receivership, but the insurance inspectors found the Life Insurance Company solvent. Its assets were scaled down by 10 per cent, it was reorganized on a joint stock basis, and a loan from the Reconstruction Finance Corporation allowed it to meet immediate redemption demands.[10]

The internal problems of the Farmers' Union and Holiday Association went unnoticed against the backdrop of more important events that summer in the farmbelt. The AAA had seized the initiative with more vigor than a federal government had ever demonstrated toward farm problems. To prevent a crushing surplus of cotton, ten of the forty million acres planted before the farm bill had passed had been plowed up by August, $100 million in benefit payments expended, and the final crop reduced from seventeen to thirteen million bales. Drouth in the Northwest made emergency measures to deal with the wheat surplus unnecessary and a production control system granting benefit payments to farmers who signed three-year reduction contracts was in effect by midsummer.[11]

Corn and hogs, the products of the areas where protest had been most intense, provided so complex a problem that Secretary

---

[9] John Chalmers, at the time president of the Iowa Farmers' Holiday Association, replied, when I asked him about these charges, "I wouldn't say they weren't true." He recalled that some of his travel expenses to Holiday meetings had been paid by the insurance company. Personal interview, October 21, 1961.

[10] Reno to J. E. Anderson, June 6, 1933; to Charles Dean, June 24, 1933; to Mary Osborn, October 3, 1933; to E. E. Kennedy, July 15, 1933; to S. W. Kaster, August 28, 1933. The company, now independent of the Farmers' Union, maintains headquarters in Des Moines. Milo Reno's grandson is its president.

[11] Wallace, 178; Schlesinger, *Coming of the New Deal*, 59-62.

Wallace saw no immediate solution. The corn crop in 1932 had been bountiful and the prices low. Farmers, therefore, fed most of their corn to livestock, gambling on the possibility of good returns for hogs and cattle the following year. Since the "hog cycle" took eighteen months to run, there was a prospect of a serious glut on the market in the winter of 1933.[12] An answer to this problem came out of the cornbelt. In early June, J. S. Russell, agricultural editor of the *Des Moines Register,* and Roswell "Bob" Garst, a large livestock producer, called together representatives of Iowa farm organizations to forward suggestions on the corn-hog problem to Washington. John Chalmers of the Iowa Holiday Association and Glen Miller from the Farmers' Union attended but withdrew in disgust. Leaders like Russell and Garst were a different breed from those who urged withholding movements and demanded cost of production. A shift of initiative from radical farm organizations to more conservative leadership, although not recognized at the time, was taking place.[13] The idea of a producers' advisory committee so appealed to Henry Wallace, searching for some solution to the corn-hog dilemma, that he decided the Department of Agriculture should organize one in each state. With the blessing of the Secretary, state representatives met in Des Moines in July to form a National Corn-Hog Committee. From this meeting came the suggestion that the Agriculture Department should buy up and slaughter millions of little pigs as a way of checking the impending oversupply. The slaughter program was announced on August 18 and through the autumn months 6.2 million little pigs and 222,000 sows were purchased and destroyed. Farmers were compensated on a sliding scale ranging from $6.00 to $9.00 per hundredweight and 100 million pounds of edible pork were distributed for relief purposes. Payments to farmers totaled about $30 million.[14]

The conferences that conceived the pig slaughter program provided Milo Reno his answer as to the direction the New Deal agricultural program was taking. Reno attended both the first

---

[12] Wallace, 179; Schlesinger, *Coming of the New Deal,* 62-63; *Wallace's Farmer and Iowa Homestead,* October 15, 1932, p. 536.

[13] *Farm Holiday News,* July 24, 1933; Korgan, "Farmers Picket the Depression," 209-11. Bob Garst achieved some further prominence when Nikita Khrushchev visited his Iowa farm in 1959. *New York Times,* September 24, 1959, p. 1.

[14] Korgan, "Farmers Picket the Depression," 211-12; Wallace, 179.

National Committee gathering in Des Moines and a livestock producers' conference in Washington the following month. What he saw ended his indecision. He believed the destruction program was indefensible, but his principal criticism was directed at the personnel who dominated the meetings and emerged as the leaders of the AAA in Iowa. Looking about the conference at Des Moines, Reno recognized officials of the Farm Bureau from most midwestern states and a host of county agents. He noted the presence of an old antagonist of his, Ralph Moyer, one of the leaders of the Farmers' Union bolters, and Dr. John D. Black of Iowa State College. Reno had been fighting this group since the days of the Corn Belt Committee and the Cedar County Cow War. "In the hog and corn conference," he wrote John Simpson, "there was an exhibition of corrupt ring politics, unequalled in any meeting held by the old group that wrecked the Hoover agricultural program." The exclusion of the Farmers' Union, the Holiday Association, and Hirth's Missouri Farmers' Association was categorized by Reno as "the most conspicuous exhibition of political ingratitude I have ever observed." The Farm Bureau, he charged, had supported Hoover, while the strength of the Democratic party in the Midwest had been in the three organizations ignored.[15] Reno took the floor at the conference in Washington to warn that unless "jingo politics" was eliminated and all groups of livestock producers equitably represented, the program would suffer the same fate as the Federal Farm Board.[16]

Milo Reno tended to personalize issues and he focused his bitter disgust and disenchantment upon one person: Secretary of Agriculture Wallace. His statements that the secretary was "about as crazy as they make them" and his program "uneconomic, unchristian and inhuman" were mild compared to later charges that Wallace was a "liar," "a disgrace to his office," and a henchman of Wall Street.[17] Reno's animosity was goaded by a near-paranoid

---

[15] July 26, 1933.

[16] Reno to William Hirth, August 14, 1933.

[17] Reno to W. H. Boreman, October 3, November 1, 1933; Reno, "Is the New Deal a Square Deal?" Radio address of May 13, 1934, quoted in White, 185-86; Kramer, 244. Henry Wallace, although he was probably unaware of the insults Reno heaped upon him, did not share this personal antagonism. Thirty years later Wallace recalled Reno with a bit of nostalgia as "a colorful, frontier type of evangelist farm leader, an orator and leader who would keep his followers in line with fist fights." Wallace to the writer, November 28, 1961.

belief that the Union and the Holiday Association had been betrayed by Henry Wallace. He delighted in referring to the Secretary's statements when he was a mere Iowa farm editor endorsing limited and controlled inflation.[18] He made caustic references to a meeting in July, 1932, where Wallace, although doubtful of the efficacy of the planned withholding movement, had suggested that farmers at a given time should burn a load of corn in their front yards as a symbolic gesture.[19]

Reno was particularly incensed by Wallace's criticism of currency inflation panaceas. On July 12, the Secretary warned of speculative price advances since May and noted that corn, oats, and wheat were selling at above production prices. A few days later these artificially inflated prices collapsed; on July 18 the bushel price of wheat fell from $1.00 to 80¢.[20] With a quick flick of logic, Reno concluded that Wallace was personally responsible and charged that this was a deliberate maneuver to lower prices and force farmers into the administration farm program.[21] In order to counter growing expansionist sentiment, Wallace warned on September 1 that inflation was a false remedy that would do nothing to adjust supplies of agricultural produce to domestic demand.[22] In Reno's words, this was "the last straw necessary to break the camel's back." In an angry attack he demanded the removal of Secretary Wallace and petitions to the President were circulated among the Holiday membership.[23] Reno was unfair to Henry Wallace. The Secretary of Agriculture was not among the

[18] *Wallace's Farmer and Iowa Homestead,* April 2, 1932, p. 3.

[19] Reno to Wallace, May 8, 1933. A copy of the minutes of the meeting of July 15, 1932, signed by E. E. Kennedy, is in the collection of Mr. George Ormsby, Wilton Junction, Iowa.

[20] Korgan, "Farmers Picket the Depression," 199-200.

[21] *Farm Holiday News,* August, 1933; Reno, "Is the New Deal a Square Deal?" quoted in White, 184.

[22] Wallace's statement was astute and its validity was confirmed by experience six months later with a limited inflation program. Currency inflation, he pointed out, might temporarily raise farm prices, making production adjustment more difficult. It would ease farm tax and mortgage burdens but, at the same time, raise prices of items farmers purchased. Cheaper dollars abroad would encourage the export of speculative farm commodities.

[23] *Farm Holiday News,* August, 1933. The paper was lagging in its printing schedule, hence the Wallace statement of September 1 was discussed in the August issue. A copy of the Wallace removal petition is in the Reno Papers and several of them are in the Official Files (Agriculture), Roosevelt Papers, Hyde Park, New York.

conservative administration advisors who fought all inflationary pressure and he wisely viewed the domestic farm problem in the broad context of the international economic crisis.

The hostility of Milo Reno did not extend beyond Wallace to the President. "I have not lost faith in President Roosevelt's good intentions," he wrote in a mid-September editorial. He had not demonstrated toward agriculture "the wisdom and consideration that we had a right to expect"—but that was due to his poor choice for a Secretary of Agriculture.[24] Even in September, with pig slaughter underway and benefit payments being made to wheat and cotton farmers, Reno harbored the belief that the President could be won back to the cost of production plan. He telegraphed Roosevelt, "Bankrupt prices of farm products and failure of farm refinancing program have destroyed farmers' confidence and hope for any relief under this administration. Sentiment for nationwide strike growing daily. Urge you act immediately for farm price stabilization based on production costs."[25]

Other voices of protest were being raised in the Midwest by the autumn of 1933. The Agricultural Adjustment Administration, after its auspicious beginnings, had stumbled. Farm prices had swerved downward after the early summer advances. The National Recovery Act, inaugurated in August, forced up prices at the grocery and implement store while the farmer's income dropped. Benefit payments were already being paid to wheat and cotton growers, but all the Agricultural Adjustment program had offered to corn-hog producers was the emergency slaughter plan to check a temporary surplus. Inflationists, vexed that the President had not yet implemented the discretionary powers granted in Title III of the farm bill, renewed their pressures. Now the arena was the executive, not the legislative, branch of the federal government. Supporting the inflationists were such friendly senators as George Norris and some unexpected converts like Representative Sam Rayburn of Texas and Senator Pat Harrison, conservative chairman of the Senate Finance Committee.[26] Governor William L. Langer of North Dakota proclaimed the farmer the "forgotten man" of the NRA and good Democrats like Clyde Herring and Charles Bryan criticized the

---

[24] MS copy in Reno Papers.

[25] Reno to Roosevelt, September 1, 1933.

[26] Schlesinger, *Coming of the New Deal,* 236.

effects of the Blue Eagle codes upon their farmer constituents. Floyd B. Olson, who had counseled patience in May, now openly demanded price fixing in place of domestic allotment. Milo Reno's disenchantment and the renewal of strike threats by the Farmers' Holiday Association came at a time when widespread dissatisfaction over the New Deal's early progress in agriculture seemed to be stirring in the Midwest.

Ten

AN AGRARIAN REVOLUTION

Gentlemen, I have called you together to inform you that the question of our buying gold is an administration policy . . . if we continued a week or so longer without my having made this move on gold, we would have had an agrarian revolution in this country [Franklin D. Roosevelt, remarks to a conference of advisers, October 29, 1933, quoted in John M. Blum, *From the Morgenthau Diaries: Years of Crisis, 1928-1938* (Boston, 1959), 72].

"Undoubtedly the agricultural situation is becoming more and more tense day by day," Henry Wallace wrote to the President on September 7.[1] A major ingredient in the increasing unrest in rural areas was the renewed pressure and activity of the Farmers' Holiday Association. Out of the atmosphere of increased disillusion following the midsummer price declines and the acceleration of inflationist pressure, Milo Reno sensed a rare opportunity. He was convinced by now that Henry Wallace and his braintrust advisors would resist cost of production price guarantees and cheaper credit for farmers. If all the dissident elements—inflationists, governors, groups dissatisfied with the operation of the National Recovery Act—could be gathered into a unified force with the militant farmers in the lead, the pressure could compel the administration to adopt a new agricultural policy.

This was the most reckless venture attempted by Milo Reno and the Farmers' Holiday Association. It was nothing less than a threat to the national administration that unless the present agricultural program was abandoned and cost of production sub-

[1] Official Files (Agriculture), Roosevelt Papers.

stituted, they must face the possibility of a debilitating farm strike. Reno grew increasingly belligerent as he realized that his confidence in the Roosevelt administration had been misplaced. This was a personal vendetta; Reno acted independently of other radical farm leaders in the autumn of 1933. John Simpson may have approved, but he was not party to Reno's schemes.[2] While Reno was trying to coordinate resistance forces, important Farmers' Union leaders in the Northwest like A. W. Ricker, editor of the *Farmers' Union Herald*, and M. W. Thatcher of the National Grain Association were touring wheat areas urging support for the AAA.[3] William Hirth wrote in the *Missouri Farmer* of October 1:

> . . . I have no patience with the recently announced intention of the FHA to stage a new "farm strike" if their every wish is not immediately granted, and this is not because some of these demands are in my opinion, unreasonable, but when the President and his advisors are doing the best they can, why not cooperate with them instead of making their gigantic tasks still more difficult? . . . Anybody who has even meagre analytical power realizes that we are passing through the most dangerous period our country has ever faced, and he who at such a time suggests "raising hell" is no friend of those whom he professes to advise and lead. . . . If there ever was a time when the last mother's son of us should use whatever common sense we possess, that time is now.

Hirth's criticism deeply hurt Milo Reno,[4] but in no way deterred him. He had a obsessive dedication and sense of purpose; he never seemed to doubt that a majority of midwestern farmers supported him and he conceived his mission as nothing short of saving the republic from destruction.

The grand design of Milo Reno and the Holiday Association was to form a powerful grass-roots lobby of inflationary political interests, dirt farmers, and a few labor and business leaders who were discontented with the National Recovery Act. In the forefront would be a group of farmbelt governors, either friends or temporary political captives of protest forces, who would bear to Washington demands for a new farm policy. Buttressing this effort would be the direct action of farmers, confined to a peaceful withholding movement at first, but if their demands were

---

[2] Reno to Simpson, October 6, 1933.

[3] *Farmers' Union Herald*, October, 1933.

[4] Reno to Hirth, October 22, 1933.

rejected it would burgeon into a full-fledged general strike of food producers.

The strategy of the Farmers' Holiday Association began to unfold with a meeting of the national executive committee on September 22. Out of this meeting, two key demands emerged. First, Reno dispatched a letter to President Roosevelt asking that he use his powers to stay all further farm executions until assistance came to farmers through currency inflation or guaranteed cost of production prices. "If you have any reason why the farmer should not be given the same consideration as banks, insurance companies, and, by edict, protected in the possession of their property . . . I would be interested in having them," he concluded.[5] The second demand was cost of production in a compelling new guise. The National Recovery Act, currently under attack in farming regions, permitted businessmen to draw up production codes, including agreements on production levels and eliminating price competition by setting a fair return on their products. Why shouldn't farmers draw up a production code just like merchants or manufacturers? The farmers' code would specify the cost of production price as the fair return.[6]

To plead the case for these demands to the Department of Agriculture, the Farm Credit Administration, and the President, a committee of three prominent Holiday leaders was sent to Washington. Announcing their coming to the President, Reno threatened, "National Farm Strike held in abeyance pending acceptance of this code." Reno confided to Simpson he had little hope for the success of the mission, ". . . consequently, I see no way of avoiding calling a strike and when this is done, no man knows what the end will be." [7]

The committee, which ventured to Washington expecting a rebuff, was surprised and perplexed by the warmth and cordiality of their reception. They had friendly interviews with the President, Secretary Wallace, George Peek (administrator of the AAA), and Henry Morgenthau, director of the Farm Credit Administration. Morgenthau informed them that any farmer threatened with immediate foreclosure should wire his office collect and he would

---

[5] Reno to Roosevelt, September 26, 1933.

[6] The author of the plan was E. E. Kennedy, secretary of the National Farmers' Union. Kennedy to Reno, September 7, 1933.

[7] *Iowa Union Farmer*, September 6, October 7, 1933; Reno to Roosevelt, September 23, 1933; to Simpson, October 6, 1933.

do everything in his power to prevent eviction. The committee reported that officials had informed them that no legal barriers prevented farmers from submitting a code of fair competition and if presented it would receive favorable consideration. "We respectfully submit," they concluded, "that for the first time in our history official Washington conceded that agriculture is entitled to receive cost of production prices." [8]

Even Milo Reno was momentarily optimistic. "My own personal judgment," he told his Holiday supporters, "is that a militant strike should be held in abeyance until it can be determined if the administration is attempting to redeem its pledges, or else simply juggling for time." [9] Nevertheless, Milo Reno had experience with previous administration promises—he was considerably chastened by what had happened in May. To a private correspondent, he wrote, "The Farmers' Holiday Association cannot possibly retain its leadership, unless it leads. . . . I regret the necessity of a strike, but in my opinion, it is unavoidable, and we must meet the issue squarely and take whatever steps are necessary even though they may, at this time be termed radical and destructive." [10]

To keep pressure constant despite hopeful omens from Washington, Reno asked state presidents to poll the Holiday membership on the question "Should there be an immediate holding action?" and on the same day, October 10, he visited Governor Clyde Herring of Iowa. Herring was either reconciled to Reno after the recriminations of the spring before or, more likely, given the unrest among farmers, he was fearful of what political power this man might command. The Governor agreed to issue an immediate invitation to fourteen governors of farm states to meet in Des Moines on October 30. The proposed agenda was drawn up by Milo Reno. The governors would hear the report of the Holiday Association committee returned from Washington, they would agree upon some action to win official approval for a farmers' NRA code, and they would plan for a general moratorium on foreclosures until cost of production was achieved.[11]

[8] E. E. Kennedy, John Bosch, H. C. Parmenter to Reno, October 10, 1933.

[9] Editorial, *Farm Holiday News,* October, 1933.

[10] Reno to Leon Vanderlyn, October 8, 1933.

[11] Reno to Herring, October 10, 1933; Herring to Reno, October 13, 1933; *New York Times,* October 16, 1933, p. 29.

The changed attitude of midwestern governors toward radical farm legislation was a good barometer of how serious the discontent in the Midwest had become and how strong pressures for a change in agricultural policies had grown. Farm-state politicians could not remain oblivious when commodity prices were falling and farm costs, thanks to the NRA, crept upward. Governor Schmedeman of Wisconsin, who had had his troubles with milk strikers, indicated his willingness to assist "in a forceful but dignified demand which will immediately put agriculture in Wisconsin on the road to restoration." Governor Floyd B. Olson noted that farm prices had not kept pace with the NRA-inspired increments to other forms of business and endorsed the plan for a farmers' production code. Governor Herring expressed his readiness to cooperate in any move to aid farm prices. Charles Bryan of Nebraska, who had been no friend of farm strikers a year earlier, declared, "I do not believe the farmers could be criticized for withholding grain from the market in order to get better prices any more than the recommendations from Washington that the public should confine their buying only to those who signed the NRA." [12]

One governor needed no encouragement. Governor William L. Langer of North Dakota declared an embargo on October 16 on all wheat sales in his state at less than cost of production prices and he asked governors of all other wheat states to join him. He stated there was a strong bipartisan demand in North Dakota for a farmers' code that would set a minimum price for grain. Reno's hand was strengthened—and very likely forced—by Langer. If Reno had doubts about calling a farm strike, they disappeared once Langer seized the initiative.[13]

The weeks before the governors' conference were a period of frenzied activity; Milo Reno delivered speeches, issued press releases, organized conferences—though no formal decision had been made, a farm strike was imminent. At Shenandoah, Iowa, Reno spoke to an audience of 5,000 and watched while farmers

---

[12] *Sioux City Journal,* October 21, 22, 1933; *New York Times,* October 21, 1933, p. 8.

[13] *Sioux City Journal,* October 17, 1933; Reno to Usher Burdick, October 6, 1933. Actually, at Langer's request, Reno sent the message to the governors asking them to support the embargo. Langer to Reno, Reno to Langer, October 17, 1933.

paddled in effigy a dummy of Secretary Wallace. Attempting to broaden the base of support, he met with a group of Omaha businessmen who passed a resolution supporting a production code for farmers; a few days later he elicited from his old friend A. F. Whitney of the Brotherhood of Railway Trainmen a pledge of "sympathetic cooperation." [14]

The culmination came on October 21 when the Farmers' Holiday Association put forth a call to the nation's farmers to withhold their produce from market until the currency had been "reflated," a cost of production code granted to agriculture, a national mortgage moratorium declared, and control of the nation's monetary system removed from the bankers.[15] Not only was this the most belligerent endeavor of the Holiday Association, it was probably the boldest attempt at political coercion ever made by a farmer protest organization in America. The public letter dispatched to all Holiday presidents simultaneous with the strike call revealed Milo Reno at his polemic best (or worst): "We have been patient and long suffering. We have been made a political football for jingo politicians, who are controlled by the money-lords of Wall Street. . . . We were promised a new deal. . . . Instead, we have the same old stacked deck and so far as the Agricultural Act is concerned, the same dealers." In answer for pleas for price guarantees, the same as were granted to other businesses, the farmers "received the Wallace hog program, which is nothing less than a brazen attempt to bribe the farmer to surrender the little independence he has left." He called upon his fellow farmers to "refuse to accept the many nostrums, the quack remedies of the brain trust." The issue was of even greater proportions. "Statements and promises" have become, he proclaimed, "mere gestures to lull the farmer to sleep that his complete enslavement may be completed. . . . The life of the Republic . . . is hanging in the balance." He warned, "We may expect some opposition from supposed farm leaders, who are satisfied with the crumbs that fall from their master's table, but I do not anticipate any opposition from the farmers and other

[14] *Sioux City Journal,* October 15, 1933; *New York Times,* October 21, 1933, pp. 1, 8; Reno to Whitney, October 20; Whitney to Reno, October 18, 1933.

[15] *Sioux City Journal,* October 20, 1933.

groups of society, who really desire the happiness and prosperity of all our people." [16]

Public reaction to the strike threat was, for the most part, one of seriousness and concern. The *Sioux City Journal*, which had over the past year considerable occasion to comment on unrest, said, "The wonder is not that the farmers at last have decided to strike but that they were able to restrain themselves until now. . . . There has never been an element of American society that endured more at the hands of the financial interests of the country." [17] George Peek, administrator of the AAA, stated in Washington: "all these people are trying to do is to save their homes. I, too, would fight to save my home. We have been warning the east for twelve years that things like this would happen unless the incomes of the farmers were increased." [18] On the other hand, Roland Jones, Omaha correspondent of the *New York Times*, disparaged the threats: "It appears to have a great deal of tub-thumping leadership, a comparatively small but zealous following and the sympathy of a good many bystanders who are inclined to agree to the justice of the complaint and to soft pedal the deplorability of the method employed to express it." [19]

What actual strength had the Farm Holiday strike movement in the autumn of 1933? Milo Reno habitually exaggerated both his own political importance and the strength of his following. Membership in the Holiday Association was small; its only preceding attempt, a year earlier, to rally masses of farmers in support of a withholding action had been a failure. The strength of the Holiday Association lay in the publicity attracted by activities, road picketing, and stopping of sales, which the organization had never sanctioned. Milo Reno was asking his followers in October to engage in actions more dangerous and make sacrifices

[16] Reno to state presidents, October 21, 1933; *New York Times*, October 22, 1933, p. 19.

[17] October 21, 1933.

[18] *Sioux City Journal*, October 21, 1933. It seems not unreasonable that the hopeful report of the Holiday committee in Washington a few weeks earlier was based on talks with Peek. While not a cost of production advocate, his orientation was toward marketing agreements rather than a reduction of production. He left the position as administrator of the AAA in December. See Schlesinger, *Coming of the New Deal*, 56-58; Gilbert C. Fite, *George N. Peek and the Fight for Farm Parity* (Norman, 1954), 254-66.

[19] October 29, 1933, Sec. 4, p. 1.

greater than he had ever requested before. To strike against inaction or to lobby with force for a legislative program was one thing, to strike against an established policy of the federal government was quite another. The projected strike was important because of another, immeasurable factor. The strength of the Holiday Association had never been tested. Neither Reno, the governors, nor the President could gauge the mood of the farmbelt well enough to know if the "movement element" that had sustained the Holiday in the past was sufficiently committed to price fixing and currency inflation to strike against the government of the United States.

The actual response to the strike call was out of proportion to the bluster of Reno's threats. Once again the Wisconsin Cooperative Milk Pool stampeded into direct action, although there was almost no coordination with the Farmers' Holiday. For the third time in a year near-anarchy reigned on the highways of the state. A picket was shot by an irate truck driver outside of Madison and at Marshfield a resisting shipper was beaten and seriously injured. One hundred cheese factories were closed and at Milan strikers dumped 10,000 pounds of milk. In northwest Iowa, some seventy-five pickets manned the old road-posts at James to the north, and Correctionville to the west of Sioux City. Four highways outside Council Bluffs were blocked and the stockyards at Omaha, never before affected by Holiday activities, reported a 50 per cent decline in truck receipts.[20] These impulsive activities were not sufficient to support the drastic demands Milo Reno was making, but they were enough to leave unresolved the question of how extensive and how damaging a farm strike might be. Moreover, this "preliminary withholding action" was not a full-fledged effort. It was designed to demonstrate graphically the discontent in the Midwest and pave the way for governors of farm states to present to Washington demands for cost of production codes and currency inflation.

The conference at Des Moines, October 30, attended by Governors Schmedeman, Olson, Langer, Berry (South Dakota), and Herring, was a remarkable testimony to the political pressure the small Farmers' Holiday Association could apply. Olson and Langer were strong supporters of price fixing; Berry, who had

[20] *New York Times,* October 23, 1933, p. 2; *Sioux City Journal,* October 24-28, 1933; *Willmar Tribune,* October 22, 31, 1933.

expressed his willingness to follow the New Deal farm program, was silent throughout the conference; Herring and Schmedeman came from the two states where protest had been strongest, a circumstance that made them more amenable to policies they could not fully endorse. The public sessions were packed with Holiday supporters and the one dissenter who made an appearance, President Charles Hearst of the Iowa Farm Bureau, was coldly received. The program adopted by the governors followed point by point the Holiday demands. The governors called for an NRA code for agriculture, to be administered by representatives of farm organizations; for cost of production price floors and revision of farm loan policies by reducing interest rates and extending coverage to 75 per cent of the value of a farm. The governors attached one supplement, a device for controlling surplus under a cost of production scheme. All producers, processors, and dealers would be licensed and to prevent flooding the market in any particular month, all food marketing would be controlled by the government. This implied that a marketing quota based upon domestic demand would be established for every farmer.[21]

The governors were sent on their way to Washington with a ringing editorial from Milo Reno. "Our strike will be called off," he pledged, "only when the program we have demanded is put into operation." Should the governors fail, "the responsibility rests directly upon the head of Franklin D. Roosevelt, because of his failure to comply with his pre-election promises. . . ."[22]

Washington was not unprepared for the arrival of the governors. Both Secretary Wallace and President Roosevelt had watched with increasing concern the deteriorating economic and political situation in the farming areas of the Midwest. Wallace had written the President a month earlier,

> Undoubtedly the farm sentiment is rapidly growing for price fixing based on cost of production because of the way in which the NRA is handling its program by apparently fiat methods.
>
> Franklin, I am very fearful of the short cut methods which so many of the farmers are now eager to have tried. It may be, however, that the NRA will soon make it necessary either to adopt price fixing in agriculture or else start inflation in the very near future. The political

[21] *Des Moines Register,* October 31, 1933; *Farm Holiday News,* October, 1933.

[22] *Farm Holiday News,* October, 1933.

temper of the dairy and livestock farmers of the middlewest at the present time is far worse than you realize.[23]

The President commented in a personal letter, "The West is seething with unrest and must have higher values to pay off their debts," and referring directly to Milo Reno's threats he told Henry Morgenthau, "I do not like to have anybody hold a pistol to my head and demand that I do something." [24] When certain of his financial advisers were reticent about the gold purchase program announced October 22 the President told them, "If we continued a week or so longer without my having made this move on gold, we would have had an agrarian revolution in this country." [25]

Belligerent threats of rebellion in farm areas were not the only factors that impelled such far-reaching decisions as gold devaluation and commodity loans to farmers in the last weeks of October, but the expressions of concern and the subsequent actions of the President and his advisors in the Agriculture Department indicate that Milo Reno and the Farmers' Holiday Association had an important (and unrecognized) influence upon federal policy.

The cabinet meeting of October 20 discussed the unrest in rural areas. Secretary Wallace hinted that the Agriculture Department would have an important announcement very soon and the President announced he would speak to the nation in a fireside chat the following Sunday.[26]

The President's fourth fireside chat, October 22, did not refer specifically to the current rural agitation, but part of the message was clearly directed to the nation's farmers: "I do not hesitate

---

[23] September 11, 1933. Official Files (Agriculture), Roosevelt Papers.

[24] Franklin D. Roosevelt to Mrs. James Roosevelt, October 28, 1933, *F.D.R., His Personal Letters, 1928-1945*, ed. Elliott Roosevelt (New York, 1947-50), I, 366; Henry Morgenthau, Farm Credit Administration Diary, quoted in Schlesinger, *Coming of the New Deal,* 66.

[25] Morgenthau, Diary, October 29, 1933, quoted in Schlesinger, *Coming of the New Deal,* 242 and John M. Blum, *From the Morgenthau Diaries: Years of Crisis, 1928-1938* (Boston, 1959), 72. The Farm Credit Administration Diary, deposited at the Roosevelt Library, Hyde Park, New York, is closed to researchers. Only the two authors cited here have been permitted access.

[26] It must be noted that the strike threats were but one manifestation of a wider rural discontent. Academic advocates of inflation such as Professors George Warren and Irving Fisher were in close contact with the President at this time. For a more extended discussion of policy-making in this period, see Blum, 50-77 and Schlesinger, *Coming of the New Deal,* 236-43.

to say in the simplest, clearest language of which I am capable, that although the prices of many products of the farm have gone up and although many farm families are better off than last year, I am not satisfied either with the amount or the extent of the rise, and to extend it is definitely a part of our policy to increase the rise and to extend it to those products which have as yet felt no benefit." He referred to the fine cooperation of cotton farmers with the adjustment program and added, "I am confident that the corn-hog farmers of the middlewest will come through in the same magnificent fashion." Most momentous was the President's announcement of his decision to authorize the Reconstruction Finance Corporation to purchase gold at prices to be determined from time to time by the President and the Secretary of the Treasury.[27]

The contemplated gold-buying program caused an immediate increase in the price of gold and beginning October 25, the President met each morning with close advisors to set the day's value. The increase in the price of gold had the effect of devaluing the dollar and was designed almost solely as a means of stabilizing commodity prices.[28] George Peek, who had spoken warmly of the Farmers' Holiday a few days earlier, now declared that the gold-buying program indicated the President was a friend of the farmer and went on, "Milo Reno is a very sincere fellow. As to his objectives, we all think the same as he does, but as to his methods, I think there is room for great difference of opinion." [29]

Three days following the fireside chat, Secretary Wallace revealed the details of a corn loan program that would directly affect every midwestern farmer. A farmer would be loaned from 30¢ to 35¢ per bushel on corn sealed and stored in bins on his premises. Interest on the loan would be 4 per cent, but if at the time of the sale the market price was lower than the combined principal and interest, a farmer could simply forfeit the sealed

---

[27] *New York Times,* October 23, 1933, p. 1; Blum, 68.

[28] This token inflationary program terminated in January when the gold price was officially pegged at $35 per ounce. It was less than successful; commodity prices declined slightly in November and December. See Schlesinger, *Coming of the New Deal,* 246; Blum, 69-75.

[29] *New York Times,* October 24, 1933, p. 1; Reno did not share Peek's optimism. He found "no ray of hope" in the proposed devaluation and he declared it was in line with policies of "money-changers who have well nigh wrecked this republic." *Sioux City Journal,* October 23, 1933.

corn in default. If the price was higher he could market the corn and keep the profit.[30] The plan was not entirely an innovation; a parallel arrangement for cotton farmers had gone into operation October 16, at which time the Commodity Credit Corporation had been created. The Agriculture Department moved rapidly to funnel loan payments into the areas of unrest. The first corn loan was paid to a Pocahontas County, Iowa farmer a month to the day after Wallace's announcement and the first check in Woodbury County was given out the following day.[31]

As a further measure to soothe the unquiet in the Midwest, Hugh Johnson, administrator of the NRA, was dispatched on a speaking tour. He told a Minneapolis audience, November 7, "Farm revolt may be useful, but it is no part of wisdom to revolt against our friends." The next day he spoke both at Des Moines and Omaha. Henry Wallace followed closely behind him for an Armistice Day talk at the Iowa capital.[32]

The advent of the five governors in Washington on November 2, wrote Henry Wallace, precipitated "one of the most interesting political thunderstorms I ever watched." [33] The emissaries talked with the President, Wallace, Peek, Morgenthau, and Harry Hopkins. Schmedeman read a telegram from his state that dairies were being blown up and the situation was "getting ugly." Herring warned that the possibilities of "an exceedingly serious farm strike are now at hand." The President assured them he was aware of dissatisfaction and Olson left the interview believing Roosevelt was favorable to price fixing.[34]

The governors' pleas were blunted by the steps that had already been taken to correct the situation that impelled their mission. The governors did not understand the policy of the Agricultural Adjustment Administration. It was pointed out to them that the proposed code for farmers was illegal since the so-called Huey Long amendment to the National Recovery Act forbade interference with free marketing rights of individual

[30] *New York Times*, October 26, 1933, p. 1.

[31] Schlesinger, *Coming of the New Deal*, 61; *Sioux City Journal*, November 25-26, 1933.

[32] *Sioux City Journal*, November 9, 11, 1933; *New York Times*, November 6, 1933, p. 1.

[33] *New Frontiers*, 56.

[34] *Sioux City Journal*, November 3, 1933; Mayer, 153-54.

farmers or laboring men. Second, it was demonstrated that parity prices would bring a higher return to farmers than cost of production prices computed at current figures. The governors began to waver; they modified their plan on the spot to request fixing of prices at parity rather than cost of production level.[35]

Secretary Wallace directed his strongest attack against the quota and licensing features of the governors' plan. Immediate setting of higher prices might diminish consumer demand. If the domestic market was reduced it would be necessary to set a definite proportional quota as to how much each farmer could sell. This would involve regimentation greater than anything contemplated under the Agricultural Adjustment Act. Wallace described the meeting:

> The Governors had been put on the spot by certain farm spellbinders who had an opening because of the increase in prices under the NRA at a time when farm products prices were dropping. Instead of trying to think the problem through, for the country as a whole, they had allowed themselves to be persuaded as to the practicability of price-fixing and they came down to Washington to put the Administration on the spot instead of themselves.
>
> Politically they were in much stronger position than we were. Their program sounded reasonable to the farmers who had been suffering most unfairly for many years. They came from five farm states; we had to take a national view. We knew we could get into a terrible mess if we attempted to go along with the Governors. If they had had something which was really practical and well-thought out, the answer might have been much different.[36]

Spurned by Henry Wallace, the governors made a last appeal to President Roosevelt. Olson was convinced the Secretary of Agriculture had influenced the President in the interval between the two meetings. Roosevelt endorsed Wallace's stand and thus wrote *finis* on the long attempt of the Farmers' Union and Holiday Association to make cost of production into law.

A considerably chastened group of governors retreated from Washington. Herring admitted he had been won over by the administration program. Olson indicated he respected the President but believed him wrong in trying to solve the farm program by voluntary, not compulsory, means. Only Langer was unrepentant. "This means the farmer is the forgotten man of this ad-

[35] Wallace, 57; *New York Times,* November 4, 1933, p. 1.

[36] Wallace, 57-58.

ministration," he complained and he recalled with contempt that "when Secretary Wallace called his assistants we found we were surrounded by professor after professor; professors everywhere." Privately he is reputed to have snarled, "We just voted one son of a bitch out of office and we can do it again!" [37]

Back in the cornbelt, Milo Reno ordered the farm strike thrown into full gear on all products. "The responsibility for whatever happens in the future," he thundered, "will rest squarely on the shoulders of the administration and Secretary Wallace in particular." [38] Reno was demanding a sacrifice that could achieve no political objective, for cost of production had now been rebuffed by both the legislative and executive branches of the federal government. To achieve these objectives now, a popularly elected executive and his chosen administrators would have to be replaced. Reno was demanding nothing short of a revolutionary undertaking.

For a few tense days strike activity flared in the old affected areas. When the governors wavered, the blockade at Sioux City grew stronger and hog receipts dropped to 10 per cent of normal. Pickets on highways were few; shippers, who by now had experience with blockade running, remained at home. Yet there was little action outside of northwest Iowa; even in the old troubled areas visiting reporters were surprised to find peace and quiet. As impulsively as it had risen, the flurry faded. This sporadic activity was all there was to show for Milo Reno's threats. The day Milo Reno called for an all-out farm strike a few lonely farmers were maintaining a roadside vigil even though an early winter snow filtered down in northwest Iowa—they were the last pickets of the Farmers' Holiday Association.[39]

Milo Reno commanded only a tiny following; 200 was the largest number involved in any protest activity in the autumn. Besides meager support, this abortive final effort faced pitfalls that had been absent a year earlier. First, even more than before, reckless violence plagued the movement; the November farm strike degenerated in a wave of unprecedented vandalism. Second, friends of the past, the Sioux City Milk Producers, opposed

[37] *Sioux City Journal,* November 5-6, 1933. Mayer, 154.

[38] Night letter to State Holiday Association presidents, November 6, 1933.

[39] *Sioux City Journal,* November 3-6, 1933; *Des Moines Tribune,* November 3, 1933; *New York Times,* November 5, 1933, Sec. 4, p. 6.

and resisted the Holiday efforts; third, there was almost unanimous public opposition that took its most virulent form in counter-organizations created to defeat the Farmers' Holiday Association.

Acts of violence occurred most frequently in the declining stages of the protest movement, between November 5 and 11. West of Sioux City a freight train crashed into an immense barricade across the tracks; when it halted 100 unidentified men broke seals on cattle cars and warned the engineer that if he proceeded a bridge ahead on the tracks would be dynamited. On the next two nights, railway bridges in Cherokee, Plymouth, Shelby, and Dakota Counties (Nebraska) were burned by unknown saboteurs. No pickets braved the cold to guard highways; the last actions of the Holiday Association were acts of sabotage carried out under cover of darkness. "Their acts of violence are regrettable," Milo Reno apologized, "the people have been admonished to carry on by peaceful picketing." [40] The last active support Milo Reno received was from the insubordinate element within the Holiday movement that he had never been able to control.

Officials of the Sioux City Milk Producers' Association, the group that had first taken to the highways the year preceding, complained the blockade was losing their members $2,000 daily; they requested police escorts to guide their trucks to market. A meeting of the Association November 9 was enlivened by a fist fight outside the doors as Holiday supporters tried to gain entry. Inside, I. W. Reck, president of the Association and a former Holiday member, countered arguments that the strike should be supported by producing a copy of *The Communist* and quoting from it an article by Henry Puro claiming that a Communist-controlled Holiday Association in Nebraska had 25,000 members and that the Communists exercised a "strong influence" over Reno.[41] By a large margin, the Milk Producers voted not to

[40] *Des Moines Tribune,* November 10, 1933; *Sioux City Journal,* November 6, 1933; *New York Times,* November 7, 1933, p. 1.

[41] The article was "The Class Struggle in the American Countryside," 547-58, the first of Puro's pleas that agrarian work should more definitely "show the face of the party." He overemphasized the party's role in the rural unrest to fortify his argument that party units should be formed in rural areas.

support the farm strike and Reck telegraphed the President, ". . . there will be no milk strike in this area as demanded by the Holiday Association. Our members are back of you in your program to aid agriculture and will not participate in any program to embarass or hinder you." [42]

The hostility of the Milk Producers was only one indication of the opposition the renewed farm strike stirred in northwest Iowa. The *Sioux City Journal* urged support of the President's recovery program and counseled this was the time to forget self-interest. Mayor Hayes, who once declared the Holiday movement was spreading across the Midwest like wildfire, complained that Sioux City was the only market blockaded and requested Governor Herring to send the National Guard to clear the highways.[43] R. F. Starzl, editor of the *Lemars Globe-Post,* once a friend of the Holiday, wrote of the Wallace corn loan program: "We don't care if Milo Reno does say you shouldn't touch any of that money. When you get a chance to get Uncle Sam's check for anywhere from $400 to $1,000, and even more, there is something wrong with you if you don't take it. . . ." [44] An Anti-Holiday Association was formed in Mills County on October 27; a few nights later 200 of its members stood guard over the Missouri River bridge at Glenwood, Iowa to escort farm trucks across to Omaha. Five hundred citizens met at Lemars, November 9 and elected H. W. Brosamle [45] chairman of a new Plymouth County Law and Order League, pledged to keep open the highways. From the League the county sheriff appointed captains for each township who would muster their forces when called upon to clear the roads. Brosamle and I. W. Reck were prime movers in forming a Woodbury County Law and Order League on November 10. Brosamle told the meeting, "Milo Reno is working only for himself and we've got to stop him and stop him now. We must stop fighting

[42] *Sioux City Journal,* November 10, 1933; I. W. Reck, personal interview, March 12, 1962.

[43] *Sioux City Journal,* November 6, 8, 1933.

[44] Quoted in *New York Times,* November 17, 1933, p. 3.

[45] The leadership of Brosamle is a fact worth noting. He was a former member of the Iowa Farmers' Union, from which he withdrew as a result of the battle with Reno over control of the Sioux City Livestock Commission House two years earlier, and he was an official of the "Progressive Iowa Farmers' Union," formed by the anti-Reno faction in Sioux City. *Iowa Union Farmer,* January 19, 1935.

among ourselves to accomplish our purpose." The new organization guaranteed to get any shipment to market and the willing Woodbury County sheriff offered to deputize and arm any of its members.[46] The Law and Order Leagues did not kill the November farm strike; they gave a push to a movement already staggering.

When Secretary Wallace criticized the Farmers' Holiday Association in his speech at Des Moines on Armistice Day, he was pouring water on a fire already extinguished. "It is true that I have never seen eye to eye with the leaders of the movement," he stated, "but when it has been suggested that the great powers of the federal government might be used in subterranean ways to disrupt the movement, I have insisted that the best way to stop this kind of ruckus was to get more money into farmers' hands. . . ." He compared the Holiday movement to the nerve of an aching tooth; it indicated where the ache was located and how serious it was. "But when you have said that, I am afraid you have said all that can be said for such a movement . . . let a few more heads be cracked, and a few more milk trucks upset, and I fear the reaction among consumers will be anything but helpful to farmers generally."[47]

A mood of calm, the first in over a year, was settling over the cornbelt. As Wallace expressed it, "the cornbelt rebellion had begun to subside. . . . The thunderstorm had cleared the air. It was possible then to explain the complicated corn-hog program . . . and to ask the help of thousands of volunteers in pushing this newest and hardest program over the top."[48] Hugh Johnson reported he had not once been heckled during his speaking tour and remarked, "If I had known the opposition was no stronger, I certainly would have remained in Washington." Wallace commented: "I looked for strikes but didn't find them," and he marveled that he had not been interrupted during his Des Moines

[46] *Sioux City Journal*, November 10-11, 1933. The leadership in the Law and Order League of I. W. Reck, president of the Milk Producers' Cooperative, and of H. W. Brosamle and the presence of milk producers in the League gives support to the interesting inference that the Law and Order League was made up of individuals of similar circumstance and in some instances, the very same men, who once were drawn to the Holiday Association.

[47] *Des Moines Register*, November 12, 1933; *Iowa Union Farmer*, November 13, 1933.

[48] *New Frontiers*, 189-90.

talk.[49] More significant than a few bridge burnings were the more than 1,000 meetings to discuss the corn-hog program attended by 178,000 Iowa farmers before December was half over.[50]

Milo Reno could not believe or acknowledge that his threats had evaporated into nothingness. "No man can call off the farm strike," he proclaimed and in a sense he did not intend, he spoke the truth. Farmers themselves by their apathy, indifference, and hostility had already called off the strike. Governor Langer lifted the wheat embargo on November 18 and the following day Walter Singler canceled the faltering milk strike in Wisconsin. A few Holiday members met at the National Farmers' Union convention at Omaha and formally declared the strike at an end.[51]

Why did the November farm strike collapse? Milo Reno never possessed actual power commensurate with the demands he was making. He was, probably unconsciously, putting up a front and so long as the extent and intensity of rural unrest in the Midwest were unknown, he was able to command attention from governors, the Department of Agriculture, and the President.

Demands for radical farm legislation probably served a useful political purpose. By contrast with plans for virtually uncontrolled inflation and arbitrary fixing of farm prices, gold revaluation and commodity loans were moderate and conservative remedies. As was so often the case in the years of the First New Deal, partial concessions were a device to calm sentiment for more drastic social and economic changes.

Two elements made up the Farmers' Holiday protest. The first was the formal organization consisting of the core who paid dues and were firmly committed to a cost of production and inflationary agricultural program. The second was a spontaneous movement element that in its farm strikes and penny auctions gave the Association what driving force it had. Reno erred in believing that those who blocked highways or interfered with foreclosures necessarily did so because they believed in the Holiday legislative program. The most important feature of the direct action movement was its singular focus on immediate and tangible economic bene-

[49] The Secretary did not know that Donald R. Murphy, Roswell Garst, and other organizers of the meeting had reserved the front rows for Farm Bureau members in what turned out to be a needless precaution against Farmers' Holiday hecklers. Donald R. Murphy, personal interview, March 15, 1962. *Sioux City Journal*, November 10, 16, 1933.

[50] *Sioux City Journal*, December 13, 1933.

[51] *Ibid.*, November 18, 20-23, 1933.

fits—better prices tomorrow or saving a neighbor's home today. All Milo Reno had to offer in November was continued struggle and sacrifice—small remuneration in contrast to the generous loans and benefit payments offered by the Agricultural Adjustment Administration.

The New Deal agricultural program sapped the strength of the rural protest movement. Yet its successes no more indicate farmers' support for its basic philosophy than picketing of highways implied commitment to cost of production. The AAA won support because it promised farmers the immediate economic assistance they so gravely needed. Evidence is convincing that more farmers favored cost of production than domestic allotment—but why press demands when the mail carrier was delivering government checks? [52] A random poll of Iowa farmers in November, 1933 revealed only 17 per cent favorable to the NRA; 37 per cent favorable to the hog reduction program; and 57 per cent in favor of corn loans (the program that promised money for farmers soonest). However, when asked if they were "in general" favorable to the President, 72 per cent replied they were.[53] In sum, farm protest waned when there appeared a hope that economic conditions would improve, regardless of the methods by which this would be achieved.

In its focus on immediate economic goals and lack of ideology, the farm protest of the thirties parallels earlier farm movements. The Granger movement, the Farmers' Alliance, and the Populist party all waned as farm income rose. Farm protest has been a transitory phenomena, spurred by immediate economic crisis; the promise alone of a return of better times has been sufficient to assuage it.

---

[52] See Fite, "Farmer Opinion and the Agricultural Adjustment Act, 1933," *Mississippi Valley Historical Review,* 656-73, especially 672-73.

[53] The poll was taken by the *Des Moines Register* and the results were reproduced in *New Republic,* LXXVII (November 29, 1933), 63-65. The Democratic party vote in the traditionally Republican strike areas of northwest Iowa and southwest Minnesota may also be interpreted as reflecting this phenomena. The Democrats again carried these counties in 1936, but in 1940, when the worst of the economic crisis was passed, all returned to the Republican fold. A report on the 1940 vote in Iowa and Minnesota submitted by Henry Wallace to the President (undated, President's Personal File 41, Roosevelt Papers) showed that the greatest percentage decline in Democratic votes was in those counties where Holiday protest had been most intense. It should be noted, however, that these returns may indicate isolationist sentiment as much as apathy toward the New Deal when better times had returned. See also Robinson, *They Voted for Roosevelt,* 89-93.

Eleven

THE CONTINUING PROTEST

We, the people, have yet the power to determine the destiny of this republic. We may not retain this power for long, but while we yet have it, we should use it to protect our country from the evils that have destroyed the nations of the past. . . [Milo Reno, "The Threat of Dictatorship," Radio address of January 5, 1935, quoted in Roland A. White, *Milo Reno, Farmers' Union Pioneer* (Iowa City, 1941), 155].

Farmer rebellion was stifled by the Agricultural Adjustment Act, but it did not entirely disappear. Occasional foreclosure sales were being halted as late as 1936. Milo Reno, embittered by failure, became increasingly extremist in his attacks upon New Deal policies and the men who administered them. The Farmers' Holiday Association remained an active organization although its membership and influence diminished. Its rival, the Farmers' National Committee for Action, continued for three years after 1933 its attempt to build a Communist front in the farmbelt.

Milo Reno, the principal sustaining force in the continuing Holiday Association, never doubted that the corn loan program and dollar devaluation were deliberate attempts to sabotage the farm protest. The events of 1933 confirmed all of his worst fears about New Deal farm policies. "The whole program of the Brain Trust is in the direction of Russian Communism, German Fascism, or Italian dictatorship," he wrote early in December.[1]

How much support did Milo Reno carry with him? Holiday conventions in 1934 and 1935, featuring headline speakers like

---

[1] Reno to George J. Fox, December 3, 1933.

Father Coughlin and Huey Long, attracted as many as 10,000. From 1934 until 1936 Reno broadcast to his Iowa supporters each Sunday over station WHO in Des Moines. The newspaper, reaching about 5,000 subscribers, took on a new commercial flare and was published regularly every two weeks. Nevertheless, total receipts of the Holiday Association in a year and a half of existence had been only $1,000, representing, at the maximum, payment or renewal of dues by 8,000 members.[2] Membership in the Iowa Farmers' Union was the clearest evidence of Reno's diminishing influence. In 1929, the organization could count 9,600 members, in 1935, as Reno's career came to an end, it claimed 3,880, in 1936, 2,466.[3] The death of John Simpson in March, 1934 isolated Reno even more for he no longer had close liaison with the National Farmers' Union.

No individual who makes political dissent his profession can admit to a withering away of his following. Reno insisted that Holiday membership was increasing in 1934 and in May of that year he envisioned a great progressive front taking shape against the New Deal. He knew better. Privately, he lamented that no organization in history had ever been able to weld farmers into a solid class conscious body and in his many speeches he looked to the past, not the future. For the rest of his life he lived in the memory of the days when farmers picketed highways, defied sheriffs, and pressured lawmakers.

Milo Reno talked too much; he became a demagogue. His weekly speeches grew repetitious as the same metaphors and the same wanton charges recurred again and again. His good arguments were lost in a welter of superficiality and name calling. Nearly any paragraph chosen at random from his speeches, such as the one that introduces this chapter, becomes a case study in the most blatant techniques of political propaganda.

---

[2] Reno to editor, *Daily Journal* (Portland, Oregon), December 1, 1933. Again these figures probably reflect predominately Iowa members. Reno was still complaining that other state units were not remitting their dues to the national. Reno to John Bosch, December 15, 1933.

[3] See *supra*, p. 33. *Farm Holiday News*, July 11, 1936. Despite its membership decline, the Iowa Farmers' Union had one compensation in the summer of 1935. Its amateur baseball team was defeating all the opposition the state of Iowa could produce. Star of the team was the pitcher, a young farm lad, son of a Farmers' Union family, who claimed he got his eye from throwing rotten tomatoes at Henry Wallace's agents. His name was Bob Feller. *Iowa Union Farmer*, May 4, September 7, 1935; White, 118.

The program of the AAA he described as a gigantic conspiracy, its aim to "build a powerful political machine" to perpetuate the administration in power. As to the individuals who carried out the program, he declared: "A group of men, who have been sarcastically termed the President's Brain Trust, are no more qualified, either by training or experience, than a group of Hottentots to solve the agricultural problem. . . ." The President of the United States was a "liar" who had broken the solemn pledges of the 1932 platform and misinformed the people when he told them food was not wantonly being destroyed. He would never grant that a surplus of agricultural products existed so long as people were hungry. Planned scarcity was "ungodly, criminal, insane, ignorant, nefarious"; the drouth of 1934 was the punishment of God visited upon those who would destroy the fruit of the land. By 1935, Reno was scourging the entire New Deal program as unconstitutional, its bureaus and commissions rapidly destroying the independence of the American people.[4]

The only part of the Union-Holiday legislative program that remained viable after 1934 was the Frazier-Lemke bill for inflationary and cheaper farm credit. For two Congressional sessions the bill was bottled up in the House Rules Committee while Representative Lemke labored to add sufficient signatures to a discharge petition to bring it to the floor. There was still considerable sympathy for the measure; by March of 1934 twenty state legislatures had memorialized Congress for its passage. In 1935, the discharge petition fell short only seven signatures of the necessary number; Lemke charged the President had used his political influence to undercut support. In 1936, after the AAA had been declared unconstitutional and Father Coughlin inspired hundreds of petitions of support, the bill came to a vote. Contrary to Lemke's predictions, the measure was conclusively laid to rest by a decisive vote of 235-142.[5]

A second measure bearing the names of these North Dakota Congressmen attempted to write into federal law an extended

---

[4] Reno to J. H. Sumner, December 16, 1933. The quotations from Reno's speeches are from the thirteen gathered in White, 121-92. Manuscript copies of nearly all of his radio addresses are in the collection of the State University of Iowa and that of George Ormsby, Wilton Junction, Iowa.

[5] *Farm Holiday News,* March 15, 1934, April 25, July 25, 1935, May 25, 1936; *National Farm News* (Washington, D.C.), June, 1936.

version of earlier state mortgage moratorium laws. A farmer-debtor could file what amounted to bankruptcy proceedings and if his creditors failed to reach agreement on debt adjustments, he might remain on the mortgaged property for five years, paying nominal rent and interest. This measure, like the earlier state moratorium laws, stirred little enthusiasm in the Farmers' Holiday Association both because of the receivership provisions and the organization's commitment to refinancing, not postponement, of mortgages. The Frazier-Lemke Moratorium bill passed Congress in June of 1934 only to be overturned a year later by a unanimous Supreme Court decision. Promptly repassed by Congress, it was sustained by the Supreme Court in 1937.[6]

The invalidation of the Agricultural Adjustment Act by the Supreme Court[7] elicited surprisingly little jubilation in the Farmers' Holiday Association. By this time, Reno was enmeshed in third party activities and he was convinced that under the Roosevelt administration only another reduction program would replace it. The Supreme Court decision, however, spurred the last crusade Reno undertook: a campaign to recover for farmers the money processors had paid out in the unconstitutional taxes. The rationale was that since depression conditions prevented processors from raising consumer prices, they had passed the processing taxes on to farmers by paying them less for their commodities. An Association for Repeal of the Processing Tax on Hogs had been set up originally by Reno in November, 1934. After 1936, the proposal was incorporated into a bill sponsored by Frazier and Lemke; it appeared periodically on the congressional calendar until the war years. This was exclusively an Iowa Farmers' Union crusade, but there is evidence it was partially supported by groups who had political interests other than the welfare of the nation's farmers. Donald W. Van Fleet, president of the Repeal Association, reported years later that since the Union had insufficient funds for the campaign, he had received

---

[6] *Farm Holiday News,* July 2, 1934; *Farmers' Union Herald,* July, 1934. See also Benedict, *Farm Policies,* 392. The Frazier-Lemke Act expired in 1949 and has never been renewed. For a discussion of the importance of this measure for farm debtors see Ernest Feder, "Farm Debt Adjustments During the Depression—the Other Side of the Coin," *Agricultural History,* XXXV (April, 1961), 78-81.

[7] United States v. Butler, *et al,* 197 U.S. 1 (1936).

aid from several outside sources including Dan Casement, president of the Farmers' Independence Council of America, an adjunct of the Liberty League.[8]

After 1934 the Holiday Association consisted of only two significant state units, those in Iowa and Minnesota. Not even a line in *Farm Holiday News* recorded the fate of the ephemeral units that once existed in Texas, Ohio, New York, or Oklahoma. South Dakota held its last state convention in June, 1934 and Wisconsin held no conventions and did not change officers. Usher Burdick, now a member of the House of Representatives, still headed a North Dakota Association which was tangled in the political embroilments William L. Langer stirred in the state between 1934 and 1936.[9] Of the tiny state units that had an existence on paper, the most tenacious was that in New Mexico. A state convention and several county meetings took place in 1934 and 1935. In the bitter senatorial campaign of 1934, Denis Chavez made the charge that Milo Reno was a "wild-eyed radical; everything he stands for is wrong." Bronson Cutting replied, "And what about Milo Reno, the president of the Farm Holiday Association, which has its membership all over the country and all over the east side of the state of New Mexico. Do you think these people are engaged in an effort to do away with everything which liberty-loving Americans stand for? Mr. Chavez says they are. I say they are not!" [10]

The most virile and the largest of all the state units was that in Minnesota. Financial returns indicated about 4,000 members in 1934 and the secretary reported that 1,887 paid dues in 1936. The Minnesota Holiday was a different organization than its Iowa counterpart and the relationship between the two was some-

---

[8] *Farm Holiday News,* November 1, 1934; White, 89. Van Fleet's statement is in Minutes, Iowa Farmers' Union Councillors, June 24, 1941, Page Hawthorne (president of Iowa Farmers' Union) Papers, Library of the State University of Iowa. See also James C. Carey, "The Farmers' Independence Council of America," *Agricultural History,* XXXV (April, 1961), 70-77.

[9] In 1934, Langer, just convicted to a jail sentence for illegally soliciting campaign funds from CWA employees, campaigned for governor while he kept troops at the state Capitol to prevent Lieutenant Governor Ole Olson, who had taken an oath as governor, from assuming office. Milo Reno toured North Dakota in support of Langer. *Farm Holiday News,* July 2, 1934.

[10] *Farm Holiday News,* July 2, November 1, December 15, 1934, April 25, 1935.

times strained.[11] John Bosch, the president (once described as resembling a typical country school teacher), was not a charismatic leader like Reno. From the beginning, the objectives of the movement had been more clearly defined. The Minnesota Holiday, for example, never viewed the farm strike as a device to force an immediate rise in farm prices. The Association enjoyed the sympathy of the dominant Farmer-Labor party and it reflected the political philosophy of that organization. The original withholding actions had been conceived as a preparation for an eventual cooperative marketing system and by 1934 the leaders had subscribed to the mildly socialistic "production for use" program advocated by Governor Olson. In short, while Reno and his Iowa followers focused on a few immediate remedial measures like cost of production and currency inflation to restore an individualistic and competitive system, the Minnesota organization espoused a more far-reaching and significant program for social change.[12] This distinction became increasingly important in 1934 and 1935 as demagogues like Father Coughlin and Huey Long lured the dissatisfied with immediate and compelling panaceas to cure all social ills.

To the left of the Minnesota Holiday Association, the Farmers' National Committee for Action pursued a lonely and independent policy. Mass support had waned with the coming of the Agricultural Adjustment Act and responding to the criticisms of the Extraordinary Party Conference of the summer of 1933, the "face of the party" was clearly shown in agrarian work; party functionaries took the lead in farmer conferences and countryside demonstrations.

The two party-affiliated newspapers, *Farmers' National Weekly* and *Producers' News* (Plentywood, Montana), organ of the

---

[11] Report of the Annual Convention, Minnesota Farmers' Holiday Association, October 9-10, 1934; National Farmers' Holiday Association, Minnesota Division, Report of the Annual Convention, October 29-30, 1936, FNCA files. John Bosch stated that Minnesota membership was "well ahead" of any other state, including Iowa, in 1935, *Farm Holiday News*, May 25, 1936.

[12] This same dichotomy seems to have existed in the Populist movement. Contrast, for example, the immediatist inflationary demands of western leaders of the movement, such as James B. Weaver, with the broader reform context of the Illinois movement.

United Farmers' League, castigated the New Deal as a "Fascist" program. Yet all, except the Communists, who opposed the New Deal were treated with greater scorn. Henry Puro labeled Reno a "social Fascist" and the *Weekly* contended he was a disguised supporter of the New Deal whose real objective was to commit farmers into slavery to the NRA and the bankers.[13]

If any spark of violence was left in the direct action movement after 1933, Communists were in large part responsible.[14] Alfred Tiala, president of the United Farmers' League, was sentenced to a six-month prison sentence after he led a small group of farmers protesting a foreclosure sale at Warsaw, Indiana in January, 1934.[15] Almost simultaneously, at Sisseton, South Dakota, seventeen UFL members, led by Julius Walstad, the state secretary, were imprisoned when they attempted to place an evicted tenant back on his farm. A permanent injunction barred further UFL activity in the county, although an attorney from the International Labor Defense won acquittal on rioting charges for all the defendants.[16] Five months later, Ella Reeve Bloor and about 150 from the Nebraska Holiday and the Unemployed League of Lincoln appeared in Loup City, Nebraska, where they urged women employed as chicken pickers by the Alliance Creamery to come out on strike. The group was attacked by a company of town vigilantes; six demonstrators were arrested and Mrs. Bloor served a short prison sentence.[17] Unlike the demonstrations of the preceding year, these demonstrations were something less than spontaneous—in all of them some prominent Communist leader was involved. They were too sporadic and too clearly identified with communism to have achieved anything. A few farmers may have been spared foreclosure, but their chief result was to stir a hornet's nest of opposition. Unlike the tolerance usually afforded Holiday demonstrations, opponents were ready to prosecute and persecute those identified with communism.

---

[13] *Producers' News* (Plentywood, Montana), May 4, 1934; *Daily Worker,* November 8, 1933; *Farmers' National Weekly,* October 28, 1933.

[14] According to my estimate, see *supra,* p. 77, note 1, in 59 demonstrations between 1934 and 1937, 22 can definitely be attributed to Communist-affiliated organizations. The number is most likely greater.

[15] *Producers' News,* February 2, 9, 16, 1934.

[16] *Ibid.,* February 23, March 9, June 21, 1934.

[17] *Farmers' National Weekly,* June 22, 1934; *Grand Island* (Nebraska) *Daily Independent,* June 16, 1934.

All of the work of the Communists was not disruption and protest. In the summer of 1933 and 1934, a "School on Wheels," a Ford truck which was home for three "teachers," office, and classroom, toured farm states. Farmers who paid $3.00 and furnished food for the faculty could study "History of Farmers' Struggles and Organizations," "Fundamentals of the Class Struggle," and "Problems of Farm Organization," Harold Ware originally proposed the schools and the advance agent in 1934 was Nathaniel Weyl, an economist in the Department of Agriculture. The school trained 100 farmers in seven states in 1934.[18]

The high point of party agrarian activities in 1933 and 1934 was the gathering of 702 delegates for the Second Farmers' National Relief Conference at Chicago in November, 1933. Unlike the conference of the preceding year, there was no attempt to mask the obvious "red" nature of the meeting. Across the assembly hall hung a blazing banner, "Workers of the World Unite." Only the Socialists and Communists accepted invitations sent to all political parties to send a spokesman; the *Farmers' National Weekly* reported the delegates "impatient at the generalities" of the Socialist representative but that "stormy applause" greeted the Communist presentation by Clarence Hathaway, editor of the *Daily Worker*.[19] The conference resolutions made no compromises with cost of production or reformist demands; delegates demanded total cancellation of mortgages, debts, and taxes and called upon farmers to enforce their rights with hunger marches, penny sales, and strikes.[20] The substance of the Chicago resolutions was captured in a bill drafted by the Central Committee of the party (a fact openly publicized). Although this was probably the most revolutionary farm legislation ever presented in Congress, it was no serious legislative measure.[21] Clarence Hath-

---

[18] *Producers' News*, June 21, 28, 1934; *Farmers' National Weekly*, July 15, 1933. Weyl's participation is interesting. He has stated that he was a member of the celebrated Ware cell in the Department of Agriculture and has identified Alger Hiss as a member. "I Was in a Communist Unit with Hiss," *U.S. News and World Report*, January 9, 1953, 22-40.

[19] *Farmers' National Weekly*, November 18, 1933.

[20] *Daily Worker*, November 18, 29, 1933.

[21] The bill was presented in two sessions of Congress, once by Rep. Terry Carpenter of Nebraska, the second time by Usher Burdick of North Dakota. An informal and untranscribed hearing on the bill was held April 8, 1935. *Farmers' National Weekly*, June 29, 1934, January 18, April 19, 1935.

away described its purpose: "The impoverished farmers cannot solve their problems under capitalism, nor through legislative enactment. But the fight for this bill can raise sharply the impoverished conditions of the mass of the farmers, it can become a rallying center for a mass movement. . . ."[22] The Communist party was still striving to build a revolutionary front, perhaps even, in Puro's words, "to prepare the American farmer for the forthcoming 'October Revolution.'" Proletarian farmers were to be linked in a common front with radical workers, not with Socialists, not with other dissenting farm organizations.

No marches, demonstrations, or rash of penny auctions, which had followed the Washington conference, recurred after the Chicago meeting. No participant could have had any illusion that this was not a Communist-controlled agrarian movement. The remaining party farm movement had no mass, militant support. Nevertheless, although their numbers were small, *some* American farmers were drawn to an open Communist program. Like the Holiday Association, the Communists could offer little in the way of immediate economic gains; the attraction for these farmers was ideological, not economic. Who were these farmer "fellow travellers"? Questionnaires completed by the delegates at Chicago revealed that they represented an alleged 109,885 farmers.[23] The largest group came from an ill-defined "dairy and general crop belt" consisting of Wisconsin, Minnesota, and all the northeastern states. Wheat farmers (218) were the largest single group of commodity producers represented; sixty-eight were corn-hog raisers. Most revealing was the information on economic status, supplied by 408 of those present at the conference. Contrary to expectations, 75 per cent of these delegates were farm owners and the percentage was even higher (80 per cent) among the red-tainted United Farmers' League representatives. Eighty-eight per cent of these farm owners carried mortgages and 60 per cent had no income at all from the preceding year's operation. A further breakdown revealed that 64.6 per cent of the delegates had a total investment in land and equipment between $5,000 and $20,000; 19.7 per cent had less than $5,000 invested and 15.7 per cent had more than $20,000. In sum, this evidence would indicate

[22] *Producers' News*, May 18, 1934.

[23] Since membership records in local units were poorly kept and estimates often nebulous, this is most likely an exaggeration.

that support for communism was greatest among hard-pressed middle-sized farmers, not tenants or farm laborers. Other data, collateral to this, indicated these were dissident and nonconforming individuals. Only 19 per cent (mostly from the wheatbelt) had signed AAA crop reduction contracts; only 149 of the 408 belonged to other farm organizations, forty-three to the Holiday Association, forty-six to the Grange, Farm Bureau, or Farmers' Union. Seventy per cent read one of the party farm papers.[24]

Of the organizations that made up the Farmers' National Committee for Action, the strongest after 1934 was the United Farmers' League. Estimates of its membership ranged from about 14,500 to 1,500.[25] The center of support, the same as the Communist party in the Midwest, was among the Finnish population in the Iron Range counties of Minnesota and the upper peninsula of Michigan. Both inside and outside the party, the UFL was regarded as a "red" organization. Henry Puro of the party Political Bureau served a time as executive secretary and the board of directors included individuals who had been Communist candidates in Nebraska and Montana.[26] Some of the schisms that rent the world Communist movement in the thirties trickled down to this tiny rural adjunct. Charles Taylor, a party member and president of the United Farmers' League, edited in his hometown of Plentywood, Sheridan County, Montana, a militant newspaper, *Producers' News,* which ministered to the needs of a small Finnish radical community in this otherwise Republican county and

[24] Anstrom, *The Communist,* 47-52. Of the nation's farm organizations, only those with some Communist affiliation (with the exception of the Socialist Southern Tenant Farmers' Union) spoke to the relief needs of the so-called poverty fourth of American farmers, a group whose existence was most graphically revealed by the agricultural census of 1935. This study would indicate that, regardless of Communist interest in their plight, these farm laborers, migrants, and inhabitants of submarginal lands were not attracted to organized protest activities. For detail on the condition of these "disadvantaged classes" see Benedict, *Farm Policies,* 357.

[25] A report of the credentials committee at the national convention of the UFL at Minneapolis (June, 1934), listed 14,496 members. Two months later John Barnett, "The United Farmers League Convention," *The Communist,* XIII (August, 1934), 810, gave the number of actual dues-payers at 1,500 with 3,000 membership books out. [Harold Ware], unpublished MS, "Farmers Emergency Relief Conference, South Dakota—1935," estimated 3,000 UFL members. FNCA files. I am inclined to accept the latter figure as accurate.

[26] *Producers' News,* July 14, 1934.

served as the official paper of the UFL. In 1933, Taylor imported several young party workers from New York City to edit the paper. Disagreements shortly resulted since Taylor's name was withdrawn from the editorial page and it was announced he was demanding a lien on the paper for his back salary. The troubles ran deeper, for in June, 1934, Henry Puro reported to the Eighth Convention of the Communist party that Taylor was an heretic who advocated revolution through existing political channels (a Farmer-Labor party) and he had refused orders of the Central Committee to change his line.[27] Comrade Taylor was little intimidated by party purges. At a shareholders' meeting in July, 1935, Taylor gained control of 80 per cent of the proxy votes and returned to the editorship in wrathful fury. *Producers' News* still bore the caption "Paper of the Oppressed and Exploited" but the words had a different meaning. "Plainer and plainer does it become as the days go by," declared a Taylor editorial, "that Leon Trotsky was and has all of the time been correct in his contention. It is Stalin and his bureaucratic henchmen who are betrayers of the world revolution. . . ."[28]

The Madison County Plan Farmers' Holiday was still alive enough to halt a few sales through 1936 and it held a state convention in March, 1934. This was the most genuine of grass-roots organizations in the party farm movement and even a year after the "re-organization," party functionaries had trouble holding some of the members in line. When the Frazier bill was discussed at the March convention a confused local reporter noted, "Many times the debates led the convention so far astray that neither the chairman or the delegates could remember what started the argument." Mother Bloor, seated in front of the chairman, "consistently coached him on what to say and how to say it, as well as suggesting the next business to be taken up."[29] Several of the last veterans of the original group at Newman Grove walked out when the convention refused to endorse the Frazier bill. The purge of the original leaders, arranged by Ware in April, 1933, was completed when Andrew Dahlsten was replaced as secretary-treasurer and the *Farmers' National Weekly* accused him of work-

[27] *Ibid.,* May 31, 1933, February 23, 1934. H[enry] Puro, "The Farmers Are Getting Ready for Revolutionary Struggles," *The Communist,* XIII (June, 1934), 574.

[28] *Producers' News,* August 30, November 1, 1935.

[29] *Grand Island Daily Independent,* March 22-24, 1934.

ing with renegade anti-Communist groups headed by Jay Lovestone of New York City, deposed party chairman.[30]

After the summer of 1933, Communist activity among farmers was a "made" movement, created largely by party representatives, and neither the press nor the leaders concealed the fact that it was atrophying. Alfred Tiala of the UFL noted after the Chicago conference the almost total failure to penetrate other farm organizations. A drive for added subscriptions to the *Weekly* fell short of its goal by over 50 per cent. The *Daily Worker* admitted "the fact that the *Farmers' National Weekly* does not now have any wide circulation in the ranks of the small and middle farmers, either reformist, left wing or unorganized, reflects the serious weakness of the entire movement." [31]

These criticisms indicated a change of tactics was again underway. Puro called for consolidating a left wing under the leadership of the United Farmers' League, and the only national convention ever held by the UFL called for a "united front" of all farm organizations to protest federal government negligence in dealing with the drouth that parched the Midwest.[32] In contrast to their usual petulance, the United Farmers' League invited the Holiday Association to join with it to fight the AAA, inaugurate mass action for more ample relief measures for farmers, and push for enactment of the party-sponsored Emergency Relief bill. Obviously, Milo Reno was not one to forget past recrimination and he did not honor the request with a reply, but a few weeks later he wrote, "the Holiday Association will not be associated with any group that has for its object the destroying of representative government and substituting either Communism or Fascism in its place." [33]

---

[30] July 13, 1934.

[31] *Farmers' National Weekly*, January 15, 1934; *Daily Worker*, March 9, 1935. The *Weekly* had 1,100 subscribers in June, 1933. Quotas for the subscription drive in 1934 were not exorbitant. North Dakota was to get 562 new subscribers; Iowa, 106; Nebraska, 464. North Dakota, the leading state, met only 54% of its quota; Iowa, 5%; Nebraska, 6%. "Farm Conference," undated MS, FNCA files. *Farmers' National Weekly*, September 6, December 16, 1934.

[32] Puro, "The Farmers Are Getting Ready for Revolutionary Struggles," *The Communist*, 577-78; Barnett, *ibid.*, 810-19; *Farmers' National Weekly*, July 6, 1934.

[33] *Farmers' National Weekly*, September 21, 1934; Milo Reno to William Moxness, quoted *ibid.*, December 14, 1934.

The Communists knew, however, that Reno did not speak for the entire Holiday membership. In October, 1934, Lem Harris sent a letter of greeting to the state convention of the Minnesota Holiday and a few weeks later he and Puro met with Bosch and other Minnesota leaders. They agreed that the UFL and Minnesota Holiday would cooperate to win increased relief for farmers and in insisting all relief be locally administered by farmers.[34] This was a beginning of a relationship between the Minnesota Holiday Association and the Farmers' National Committee for Action that grew increasingly friendly. John Bosch was no Communist sympathizer nor was he ever in any way controlled by party leaders, but he looked upon the front organizations as genuine farmer associations and envisaged cooperation on a limited scale as beneficial.

The new emphasis on a united front did not, as yet, mean any ideological compromise by the Communists. Communist attacks upon the Holiday program and its leaders (with the exception of Bosch) increased. They continued to press for enactment of their Emergency Relief bill.[35] This, in short, was a call for a "united front" on Communist terms. Considerations both practical and tactical underlay this changed approach in late 1934. The drouth had brought serious economic hardship to farming regions and among a lower strata of tenants and small owners (like the Okies who fled to California) there was need for more generous public relief. Tactically, the Communist party was slowly turning away from the fruitless policy of isolation from the American left. The timing was judicious for the emergence of groups headed by Father Coughlin and Huey Long as the spearhead of opposition to the New Deal threatened a division and splintering of progressive forces. Early in 1935, the Central Committee of the party urged members to look to the Farmer-Labor party of Minnesota, to "seize upon and head up" the rank and file movement growing there. It called for a "manifold increase" in revolutionary work in the Minnesota Federation of Labor, the Farmers' Union, and the Holiday Association.[36] The new amity with the Minnesota Holiday, therefore, was part of a greater design, rendered more significant because the Holiday was the smallest and most vulner-

[34] *Farmers' National Weekly*, October 12, 26, November 9, 1934.

[35] *Ibid.*, January 11, 18, 25, 1935.

[36] *Daily Worker*, February 7, 1935.

able of the groups mentioned by the Central Committee. Communists still talked of a "rank and file" movement, i.e., a "united front from the bottom" but the staging ground was changed from the isolated vantage points outside any existing organizations to a point at least on the threshold of the traditional American left.[37] The new strategy was to penetrate, not oppose, other protesting organizations.

As a further attempt to use the drouth issue to build a rank and file movement uniting Communists and members of other radical farm organizations, party leaders organized a Farmers' Emergency Relief Conference that met in Sioux Falls, South Dakota, March 22-25, 1935. "We, the Communists, have very important tasks in connection with the preparation for this conference," Henry Puro announced. It had to be demonstrated that the pauperization of the great majority of the farmers was due to the New Deal and that the farm masses were in danger of being misled by Long, Townsend, and Reno.[38] Swinging back to a more pragmatic approach, the militancy of the Chicago resolutions was lacking as the conference called for increased drouth relief, immediate repeal of the AAA, and approval of the party-sponsored farm bill. Another resolution called for replacing the Farmer-Labor party of Minnesota with a "real United Labor party."[39] This standard, a more plausible one for attracting broadly based support, did not win the attention of groups outside the Communist farm movement that the leaders had hoped for. Out of 405 who attended the conference, 181 were members of the UFL and Madison County Plan.[40] Nevertheless, the Reno Holiday Association was better represented than at Chicago; 98 members were present along with 65 Farmers' Union members.[41]

[37] It may be observed that Communist party attitudes toward radical farmer organizations paralleled almost precisely those toward the Socialist party. See Howe and Coser, 231.

[38] *Daily Worker*, February 7, 1935.

[39] *Farmers' National Weekly*, April 5, 1935. The resolutions on the Farmer-Labor party seem more uncompromising than the earlier Central Committee directive would call for.

[40] UFL: 151; Madison County (Nebraska) Farmers' Holiday: 30.

[41] This information is from an unpublished study, paralleling that made at the earlier Chicago convention, "Farmers' Emergency Relief Conference, South Dakota—1935," in the files of the FNCA, which Mr. Harris informs me was written by Harold Ware. The study again inquired into the owner-

The Sioux Falls Conference was a preliminary to the most ambitious attempt of the Communists at direct frontal assault on the Farmers' Holiday Association. Holiday members attending the Sioux Falls Conference requested that the resolutions there be presented to the Holiday at its annual convention at Des Moines, April 27, 1935. Milo Reno surprisingly accepted, although he refused to allow Lem Harris to act as spokesman.[42] The April 26 issue of *Farmers' National Weekly* bore the headline "Greetings to the Holiday Convention" and to capitalize upon what was considered a serious and growing rift in the membership, the paper dwelled at length upon the reactionary qualities of Huey Long and Father Coughlin (both of whom were believed to be speaking at the convention).

Communists could never have hoped to carry the Holiday Association by storm; at best they might siphon off some progressive elements who could not follow Milo Reno into a third party effort in alliance with Huey Long or Father Coughlin. The Holiday convention of 1935, at which Senator Long was the featured speaker, was the only time the Farmers' National Committee for Action leaders ever confronted Reno directly. It was an encounter worth detailing. Reno was absent and John Bosch presided at the business session in which a representative of the Sioux Falls Conference rose to read the conference's resolutions. Midway in the presentation, the door flew open and Reno entered. Without waiting for recognition he launched into a hot-tempered attack on Lem Harris, the *Farmers' National Weekly*, Russia, and Communists in general. When a Nebraska delegate rose to protest, Reno silenced him by calling him "Lem Harris puke." For affirmation, Reno called on F. C. Crocker of Nebraska, who had battled the Madison County Plan at the Lincoln march two years earlier; he responded with a belligerent attack upon

ship status of the delegates and the results confirmed the earlier inquiry. Of 283 farmer respondents, 169 were property holders (141 mortgaged) and 87 were tenants. (For 27, no report on tenure was given.) Outside these categories, 47 agricultural laborers and 95 "others" attended. Although this study used a different test of economic level (total income rather than total investment), the fact that 82% of the delegates had incomes less than $1,000, 16% between $1,000 and $4,000 and 2% above $4,000 would indicate greater success in attracting poor farmers than at Chicago.

[42] FNCA to Reno, April 12, 1935; Reno to FNCA, April 18, 1935, quoted in *Farmers' National Weekly,* April 26, 1935.

"Communistic Jews." The Sioux Falls program was voted down by a large majority.[43]

How far this effort at splitting and penetrating the Holiday Association might have proceeded is unknown, for the summer of 1935 was one of the vital turning points in the history of the international Communist movement. Communism had been floundering for two years, still voicing innuendoes of militancy and impending revolution, yet clearly troubled by the rising tide of fascism: the anti-semitism of Hitler, the Saar plebiscite, the adventurousness of Mussolini. Meanwhile, as events of the summer of 1935 would show, national Communist movements were being more firmly cemented into adjuncts of Stalin's foreign policy. Prompted by threats of fascism, Russia had forgotten past accusations to creep back into the League of Nations and in May, 1935, the Soviet Union concluded a *realpolitik* mutual defense pact with France. Georgi Dimitrov of the Bulgarian party, speaking at the Seventh Party Congress in Moscow, August, 1935, declared that Communists had underestimated the power of fascism and to halt its advance nothing short of a union of all antifascist forces, be they socialist or parliamentary reformers, was urgently necessary. Proceeding with a country by country analysis, he made detailed comments on the United States. He warned of fascism disguised as radicalism and cited Huey Long's Share-the-Wealth Society as an illustration. To combat menaces of this kind, he urged American Communists to take the lead in the formation of a Workers' and Farmers' party, neither Socialist nor Communist but "developing the most wide-spread movement." Among those present who cheered the advent of the new Popular Front were Earl Browder, general secretary of the American party, and Lem Harris, executive secretary of the Farmers' National Committee for Action.[44]

Hardly had he returned to America when Browder issued a call for the formation of a Farmer-Labor party in which Communists would play "only a small part." With a heedless disregard for consistency, the Farmers' National Committee for Action enthusiastically adopted the new Popular Front. The Emergency Re-

[43] *Farmers' National Weekly,* May 3, 1935.

[44] Arthur M. Schlesinger, Jr., *The Politics of Upheaval* (Boston, 1960), 564-65; Howe and Coser, 319-22; Lem Harris, personal interview, April 11, 1962.

lief bill disappeared from the columns of the *Farmers' National Weekly*. Delegates from the FNCA tried without success to merge with the Minnesota Holiday, but despite the failure the newspaper commented, "However if past prejudice and friction prevented the delegates from seeing clearly the urgent need of unity this by no means should block the unification program." [45] Fusion of the United Farmers' League into the Minnesota and North Dakota Holiday Associations proceeded nonetheless. Although there was no formal sanction from either of the state Holiday units, in county after county local units amalgamated. As late as September, the Frazier-Lemke bill was still being criticized, but on December 20 a UFL official, Henry J. Correll, wrote, "Then there are those who fail to see any good whatsoever in it, claiming it to be of no value to anyone, other than the mortgage holder, that it is a bankers' bill, that it does not contain one solitary provision of benefit to the debt ridden and impoverished farmer." To whom did the writer refer? At the August convention of the North Dakota Holiday, Henry J. Correll had condemned Frazier-Lemke as a bill "aiding mortgage holders." [46] When the AAA was declared unconstitutional, the *Weekly* responded, "We were never for it, but it has been overthrown by reactionaries of the Supreme Court. But we must fight to keep the small allowance and pittance the A.A.A. allowed." [47] By January the paper even had kind words for cost of production. Usher Burdick was hailed for his condemnation of Mussolini; John Bosch appeared as a columnist in the *Weekly* and of Milo Reno the paper noted only that he "underestimates the danger of Fascism." [48] A new legislative program announced in January called for increased relief, cost of production, abolition of acreage restrictions, and the Frazier-Lemke bill; in short, the Farmers' Holiday Association program.[49]

The major architect of the Communist party agrarian program did not live to share this new spirit of cooperation. Harold Ware was killed in an auto accident near Gettysburg, Pennsylvania on August 13, 1935. The brief obituary in *Farmers' National Weekly*

[45] November 1, 1935.

[46] *Farmers' National Weekly*, August 2, 1935.

[47] January 10, 1936.

[48] *Farmers' National Weekly*, October 18, December 20, 27, 1935.

[49] *Ibid.*, January 24, 1936.

noted, "Despite his memorable work he was known to but few of us . . . he was so unassuming and retiring that only those with whom he came into intimate contact knew of his work."[50]

Tangential to its political activity in the Midwest, the Communist party utilized a different and unique method of publicizing its position on the farm question in the spring of 1936. On March 14, the Federal Theatre Project of the Work Projects Administration presented in the Biltmore Theatre, New York, the first of its "Living Newspaper" plays, *Triple-A Plowed Under*. Ostensibly, this new dramatic media was to present "an authoritative dramatic treatment" of current problems based upon actual source material. *Triple-A Plowed Under* fell short of this objective: it was an undisguised Communist interpretation of the agricultural situation. In scene 4, Milo Reno was depicted signing an agreement with a committee of Commission merchants to call off the farm strike. Scene 9 consisted of a long speech delivered by an actor representing Lem Harris at the Farmers' National Relief Conference in December, 1932. The AAA was condemned for reducing food production and raising commodity prices for the benefit of wealthy speculators. The play concluded with midwestern farmers joining with laborers and consumers to form a Farmer-Labor party and adopting as their platform a program identical to that presented in the *Farmers' National Weekly* of February 14, 1936. The play continued in New York for several months and was presented also in Chicago, San Francisco, Cleveland, and Los Angeles.[51]

Milo Reno may not have known of the dramatic shift in Communist policy, but his major concern in late 1935 was to unify all dissident elements for a forthcoming political showdown against the New Deal. On December 23 he wrote Lem Harris,

> I am ready to sit down and discuss the farm and other problems with you if the object of such discussion is to seriously consider our deplorable situation from all angles.
>
> Frankly, I have very little confidence in yourself and some of those associated with you, for the simple reason, that you have, without investigation or knowledge, attacked the Farmers' Holiday Association and its leadership. It has been apparent to me that your methods were

[50] *Ibid.*, August 23, 1935.

[51] Federal Theatre Project, *Federal Theatre Plays: 1. Triple-A Plowed Under, by the Staff of the Living Newspaper, 2. Power, A Living Newspaper by Arthur Arent, 3. Spirochete, A History* (New York, 1938), viii, x, 3-57.

reprehensible, in fact, you seemingly proposed to build by attacking others instead of organizing your own group. I am saying this to you, Mr. Harris, because I want you to understand exactly how I have felt as to your tactics and will continue to feel until I am convinced that I am wrong. I think, perhaps we could come to a better understanding, if we could discuss things frankly, without the handicap of either defending or attacking our organizations.[52]

Unfortunately, the projected meeting never materialized. Reno, apparently, was interested in no more than exploring possibilities of cooperation, for he refused the request of Harris that he endorse the new legislative program of the FNCA although it was virtually a carbon copy of that of the Holiday Association.[53]

Those who remained loyal to organizations of agrarian protest until 1935 were a small band of stalwarts, drawn not by rational concerns for their economic welfare, but because of ideological commitment to cost of production, production for use, or the Communist party farm program. Their numbers were few and their programs were lost issues. Yet they had significance in a larger context. The depth of dissatisfaction with the New Deal measures was unknown. A host of alternatives competed for the potential loyalty of the discontented: the Social Justice movement of Father Charles Coughlin, the Communist Popular Front, the budding Farmer-Labor political movement. In all of these, the organizations of farm protest surviving from the battles of 1933 played a part.

---

[52] FNCA files.

[53] Harris to Reno, January 29, 1936; Reno to Harris, February 11, 1936. FNCA files.

Twelve

THIRD PARTY ACTIVITIES

Incidentally, Batcheller, I have lost all hope in correcting anything through either of the old parties and while I do not wish to take any prominent part in the organization of a third party, I believe it is our only salvation and the big problem connected with such a movement is to crystallize all of the many objecting groups into one powerful, militant force [Milo Reno to J. W. Batcheller, Gayville, S.D., December 23, 1933].

In the initial issue of *Farm Holiday News*, one month before the inauguration of President Roosevelt, Milo Reno suggested that a new and independent political party might be the only means to save the republic from the destruction he feared.[1] As Reno's disenchantment soured to bitter opposition, this alternative became the central objective of the continuing Farmers' Holiday Association he headed. Reno had no personal political ambitions, but, as in the drive for inflation and cost of production in October, 1933, he conceived the Holiday as a rallying point where disjointed protest elements might unite around a common standard. His activities constitute a small but important thread that weaves through the developing political dissent of the early New Deal years.

The popularity of Roosevelt and the boundless faith that something was being achieved sometimes obscures the substantial sentiment on the left that the New Deal had not gone far enough. John Simpson reported late in 1933 that he knew of some sixty-

[1] February 12, 1933.

eight third party movements far enough advanced to have a name and a headquarters.[2] Most of these were crank outfits, but there were sources of deeper disaffection. Legislative benefits had not extended far enough that segments like the old folks, some small businessmen, and farmers were immune to the appeal of charismatic leaders who promised easy wealth-distributing panaceas. One such leader, Huey Long, excited enough support that a poll in 1935 predicted he might win 2.75 million presidential votes.[3]

The most rational of the emerging left wing movements was the League for Independent Political Action founded in 1929. Its leaders were a group of Socialists and intellectuals like Congressman Thomas Amlie of Wisconsin, Rev. Howard Y. Williams of St. Paul, Professor Paul Douglas of the University of Chicago, and Alfred Bingham, Nathan Fine, and Selden Rodman of New York City, the latter editor of the organization's journal, *Common Sense.* John Dewey was the chairman. Unable to consolidate the various splinter third parties in 1932, the group had endorsed Norman Thomas but congressional candidates like Marion Zioncheck of Washington, Amlie in Wisconsin, and Elmer Thomas in Oklahoma, running with League support, had been elected. To chart a course independent both of the Socialist party and New Deal reformism, the organization met at Chicago in September, 1933. The Farmer Labor Political Federation established there broadened the base of the organization by adopting a political program that had more appeal to agrarian radicals of the Midwest. Congressman Amlie emerged as the dominant figure and the platform endorsed cost of production prices for farmers and, in calling for a cooperative commonwealth, echoed the political philosophy of Governor Floyd B. Olson. This was a half step toward a third party; the organization was committed to a definite program but remained flexible in terms of candidates and future bargaining power. Milo Reno was one of the signers of the conference call.[4]

---

[2] Simpson to Norman Lermond, n.d., enclosed in Simpson to Reno, November 6, 1933.

[3] Schlesinger, *Politics of Upheaval,* 251-52.

[4] Donald R. McCoy, *Angry Voices: Left of Center Politics in the New Deal Era* (Lawrence, Kansas, 1958), 4-25, 30-42. The "Four Year Presidential Plan" adopted in 1932 by the League for Independent Political Action called for immediate relief for the working force from depression by the expenditure of three to five billion dollars for public works; soak the rich

The new Farmer Labor Political Federation endorsed the strike efforts of the Farmers' Holiday Association in October and November. Amlie declared publicly that the farmers were making use of the only weapon that was worth a damn and predicted the strike would give birth to an agrarian third party movement. Howard Y. Williams, the national organizer, wrote Reno that the Federation was ready to cooperate in every way with the Holiday Association.[5]

The political initiative and encouragement of the Farmer Labor Political Federation came at a propitious time for Milo Reno and the Farmers' Holiday Association. Reno, who had received his political education years earlier in the Greenback and Populist parties, was the first of the radical leaders who had endorsed Roosevelt to call specifically for a new political party. Even while he was preparing the last desperate and abortive crusade against the course the brain trust had set for agriculture, he was writing to Ernest Lundeen, "It will be absolutely impossible to ever clean up the stinking mess that has been made of affairs except through a third party movement."[6] When the strike movement faltered and collapsed, Reno was determined to forge ahead. Now is the psychological time to launch a third party, he wrote John Simpson. To A. F. Whitney of the Brotherhood of Railway Trainmen, a friend with whom he often exchanged confidences, Reno suggested a meeting of a few reputable leaders in the near future to plan a conference at which a new People's party would be formed. Whitney did not share all Reno's disillusion, but he agreed that the Roosevelt brand of liberalism had thus far proven inadequate. Rather than an independent course, he suggested the Farmer Labor Political Federation should form the nucleus for their efforts.[7]

Reno was thinking of an independent movement and apparently had in mind something more than the gradualism of the Farmer Labor Political Federation; nevertheless, he was willing to cooperate. Williams talked with Reno and Simpson during the

---

taxes; workers and old age insurance; prohibition against child labor; a six-hour working day; 25 per cent reduction in tariffs; prohibition of military conscription; and immediate recognition of the U.S.S.R.

[5] Amlie to Reno, October 21, 1933; Williams to Reno, October 22, 1933.

[6] October 8, 1933.

[7] Reno to Whitney, November 6, 1933; Whitney to Reno, November 9, 1933.

Farmers' Union convention at Omaha in November. A few days later he was in Des Moines planning with Holiday leaders a Farmer-Labor party for Iowa. Reno attended the conference, but did not sign the conference call.[8] Despite these auspicious early contacts an ideological chasm separated an evangelical property-oriented Populist like Milo Reno, for whom currency inflation and guaranteed prices were the ultimate objectives of reform, from intellectual, urban socialists who advocated a cooperative commonwealth. These differences were made manifest when Reno appeared at Cooper Union, New York City in January, 1934 in a symposium with Whitney and Amlie on the issue "Are We Ready for a Third Party for the People?" Like William Jennings Bryan before him, he took New York something less than by storm. He most likely embarrassed the more cautious leaders of the Farmer Labor Political Federation by boldly suggesting that the newly elected mayor of New York City, Fiorello LaGuardia, should be the third party standard bearer in 1936. In his address, he lectured Manhattan socialists and intellectuals on the sanctity of the Constitution, insisting it was not only the safeguard of individual rights against bureaucratic infringement but that it was an adequate foundation for all necessary social reforms.[9] One startled listener wrote to Williams a few days later,

> That is the first time I have ever seen or heard Milo Reno. I had hardly expected anything like that. And yet, I suppose it is only natural. It's the good old American pietistic individualism based upon a philosophy of natural rights. He apparently thinks these natural rights to be imbedded in existing law and all that is necessary is to force the government to administer the law properly. It's a simple view of the social process and very attractive, but also woefully misleading. . . .
>
> In a way that meeting was rather startling and perhaps a bit ludicrous. Just think—an old line American farmer, individualistic, pious, land-conscious, trying to preach to a flock of New York Jews about Christ, Judaism, and land economics. I can imagine that there was very little if any raprochement of minds except possibly in bits of humor and some of his over-simplified illustrations. Farmers are discouraging from the point of view of their individualism. . . .[10]

[8] Reno to Williams, November 10, 1933.

[9] *New York Times*, January 4, 1934, p. 6; *New York Evening Journal*, January 5, 1934.

[10] Harvey Pinney to Howard Y. Williams, January 9, 1934; Howard Y. Williams Papers, State Historical Society of Minnesota, St. Paul.

This cleavage underscored the ambiguity in the forces that clustered together in the anti–New Deal left in 1934 and 1935. Intertwined in the ranks were those who advocated such social reforms as production for use or a cooperative commonwealth as well as fanatics who searched for some panacea that would move backward, not forward, to re-create a lost individualism. Both competed for the same constituency. Milo Reno, Charles Coughlin, and Huey Long were considered the heirs of the progressive tradition just as much as John Dewey and the editors of *Common Sense.*

Milo Reno did not sever contacts, but his ideological home was not in the Farmer Labor Political Federation. This was not true of all of the Farmers' Holiday Association. The large Minnesota unit, closely associated with the Farmer-Labor party, looked to Floyd B. Olson as a potential third party candidate. National headquarters of the Farmer Labor Political Federation was St. Paul and John Bosch, president of the Minnesota Holiday, was a key leader of the midwestern wing of the Federation and of its successor, the American Commonwealth Federation.

The year 1934 was an incubation period for a series of protest movements differing sharply from the refined and gradual program of the Farmer Labor Political Federation. Senator Huey Long had cut a fiery swath through the United States Senate during his two years of membership. A Roosevelt backer in 1932, he fought the administration on the National Recovery Act and inflation; the estrangement was completed in late 1933 when the federal government renewed internal revenue investigations of some of the Long machine's business activities in Louisiana. With much fanfare, Long announced the formation of a National Share-the-Wealth Society in January, 1934; he was probably already looking to the 1936 campaign, but he discreetly avoided bold moves in this direction until a year later. Father Charles E. Coughlin was still telling his vast Sunday afternoon radio audience early in 1934 that the New Deal was Christ's deal, but given Roosevelt's attitude on silver and the development of the farm policy, the program was rapidly losing its sanctity. Coughlin measured the situation until November, 1934 when he announced the formation of the National Union for Social Justice based upon a sixteen-point program that included a series of inflationary panaceas and cost of production prices for farmers. A third movement was that of praying, hymn-singing old folks who, in August, 1934, began organizing into Townsend Clubs in nearly every state. Dr. Francis Townsend, despite his mild manners and his determination to avoid party politics, had much the same kind of messianic appeal as Long or Coughlin; his plan for a monthly $200 bonus to every person over sixty-five was a slightly more rational technique of sharing the wealth. For all of these movements, 1934 was a formative period in which leaders took their estimate of the New Deal's strength and their own potential following and they determined, almost simultaneously, to risk a test of political strength. Despite similarities in appeal and potential supporters there was little coordination between the groups. The irrational left suffered a plethora of prima donnas; all these aspiring movements rested upon the magnetic appeal of some charismatic leader and secondarily upon an inclusive Utopian-type reform program.[11]

Milo Reno watched with immense satisfaction the growth of these dissenting movements which seemed to vindicate his hope that the common people were rising against the greedy financial interests and their bureaucratic allies. "The movement is in the air," he proclaimed optimistically, "it permeates every group of society. There will be an independent political party, broad and comprehensive enough in its platform to embrace every group of American citizens that is dissatisfied with the present conditions. . . ."[12] Reno conceived his role as a John the Baptist of protest: to prepare the way by building mass support and trying to draw the various groups together into an effective political coalition. He was never in the inner councils of any political group, but he had an important (and overlooked) influence. The vestigial Farmers' Holiday was a potential avenue to farmer support in the Midwest; although Reno's support had withered, he was still the most important spokesman for political disaffection within the Farmers' Union and could be regarded as the voice of a hypothetical group of dissatisfied family farmers. Both Charles E. Coughlin and Huey Long observed Milo Reno with consider-

---

[11] For a brilliant and more detailed description of the Long, Coughlin, and Townsend movements, see Schlesinger, *Politics of Upheaval,* 16-68.

[12] "Let the People Rule," Radio address of May 20, 1934, quoted in White, 163.

able interest and weighed carefully how much political support he could command.

Milo Reno had endorsed a third party at least a year before Father Coughlin even dared suggest so bold a move. Other than this, the arguments of the two men follow a remarkably parallel pattern. Like Reno, Coughlin considered the farm bill the most unconscionable of New Deal measures. Coughlin never accepted the AAA although (like Reno) he originally lauded the NRA and gold devaluation. Again like Reno, the radio priest in 1934 regarded the brain trust as the villains of the peace, and although he attacked Tugwell, Ezekiel, and Wallace, he refrained from criticizing the President. Both Coughlin and Reno attacked the President by name early in 1935. Parallelism is no evidence of influence and the likelihood is greater that Reno followed Coughlin's lead, but in two particulars, the Social Justice crusade bore some imprint of the farm protest which Reno led. One of the sixteen points in the November program was cost of production prices for farmers—a demand unique with the Farmers' Union and Holiday Association. Second only to Coughlin's determined drive against American membership in the World Court was his forceful campaign in 1936 to bring the Frazier-Lemke refinance bill, a key demand of Milo Reno, to a vote in the House of Representatives. When it was defeated he began organizing state units for the Union Party campaign of 1936.[13]

The high point in the liaison of Coughlin with the Farmers' Holiday Association was his appearance as featured speaker at the convention of May 3, 1934. In one of the most successful public appearances of his career, Father Coughlin asked 10,000 cheering patrons in the Des Moines coliseum, "Who is it, Tugwell or Christ?" Aiming his attack at the one member of the brain trust least deserving of left wing criticism, he declared, "Starve the Chinese and the jobless of this country in the name of Tugwell and you will crucify Christ again." He flayed Tugwell as one who favored the concentration of power in the hands of the "high priests of gold" and he attacked the Federal Reserve System, declaring he would never deposit money in a bank associated

[13] Schlesinger, *Politics of Upheaval*, 554-55; Edward C. Blackorby, "William Lemke: Agrarian Radical and Union Party Presidential Candidate," *Mississippi Valley Historical Review*, XLIX (June, 1962), 74.

with the system—all words that would fall on receptive ears in a Farmers' Holiday audience.[14]

With the appearance of the Townsend movement in the field in August and the formation of the League for Social Justice on November 11, Milo Reno stepped up the tempo of his third party agitation. On November 25 he praised the Townsend and Coughlin movements, predicting their growth and development. ". . . we are not going much further in the direction that we have been travelling since 1919," he declared. A few weeks later he ended a speech on "Religious and Civic Ideals" with this peroration:

> There is an agitation in the minds of our people that is healthy. Men and women are no longer controlled by blind partisan prejudice. They are commencing to think and determine for themselves. The millions who are supporting Father Coughlin's sixteen points and the Townsend old age pension plan are evidence of a great moral awakening to the absurd situation that now prevails. If this moral awakening is properly directed, directed in harmony with the principles of Jesus of Nazareth, in harmony with the divine principles of the Declaration of Independence, this nation may become as God intended, when the tabernacle or authority of God will dwell among men. . . .[15]

Milo Reno was determined that the "great moral awakening" should be properly directed. During the spring of 1935 he was hard at work on an ambitious plan to unite in a common front the Townsend, Long, and Coughlin movements. To this end he invited Huey Long, Floyd B. Olson, Father Coughlin, Dr. Francis Townsend, and Upton Sinclair to attend the Third Annual Farmers' Holiday convention. There was no basic difference, he insisted, between the Farmer-Labor, Social Justice, and Share-the-Wealth programs. ". . . it would be a crime against the Republic and the American people, if these different leaders do not come to an understanding and go into the campaign of '36 a united party," he exclaimed.[16]

Reno went to great pains to disavow that the purpose of the forthcoming convention was to launch a third party movement—so many pains, in fact, that he proved the point he was attempting to deny. "The Farmers' Holiday Association is *not* a political organization, and, so far as I know, has no ambition to sponsor a third party," he averred, but in the same breath went on to say,

---

[14] *Farm Holiday News,* May 15, 1934; *Des Moines Register,* May 4, 1934.

[15] February 10, 1935, quoted in White, 138.

[16] MS of radio address of April 21, 1935.

"I believe that ninety-nine percent will support a progressive ticket." If any doubt remained that his was a design for a third party movement, he continued,

> In my opinion, it will be a third party that will have the support of the rank and file of our citizens. Maybe history will repeat itself. Previous to the civil war, we had many dissatisfied groups, all led by more or less ambitious leaders and for a time it seemed an impossible task to consolidate the opposition, in fact, Harriet Beecher Stowe's *Uncle Tom's Cabin* did more to arouse the moral sense of the people to the crime of chattel slavery than any other one thing. Maybe that is what we need and will have to have at this time.[17]

Like so many of Reno's political ventures, this one was premature and overly ambitious. He misread the nature of the irrational protest movements and failed to detect the cynical drives for power that masqueraded behind the social nostrums of some of the aspiring leaders. Yet in terms of political strategy, Reno was correct. If these messianic movements could have found a common ground and a common leader, something more than the abortive Lemke campaign of 1936 might have resulted. As it was, Reno's efforts only played into the hands of the most opportunistic of the political knights-errant, Senator Huey Long.

Only the Louisiana Senator responded to Reno's invitation to the convention; Father Coughlin publicly denied he was associated with Reno's plans. The Farmers' Holiday convention fit in well with Long's developing political plans; already he was preparing to test his political strength against Roosevelt in the primaries in half a dozen states.[18] In preparation for Long's coming, the Farmers' Holiday leader treated his radio listeners to a long panegyric on the Senator's accomplishments, but he stopped short of mentioning him as a possible presidential candidate.[19] The convention of April 27 was a Huey Long show. Townsend Plan representatives present refused to sit on the platform. As with Coughlin a year before, the fairgrounds coliseum was filled with 10,000 sympathetic listeners who heard Reno introduce Long as "the hero whom God in his goodness has vouchsafed to his children to save us from Roosevelt, Wallace, Tugwell and the rest of the traitors." Long staged a grotesque performance. Munching peanuts through the course of his speech, he quoted

---

[17] *Ibid.*

[18] *New York Times*, April 27, p. 9; April 9, 1935, p. 11.

[19] MS of radio address of April 21, 1935.

statistics on wealth concentration, invoked Scriptural writ, and at the end called on all those who believed the wealth should be shared to hold up their hands. Everyone did.[20]

To augment further the unity of protest groups, Reno led the Holiday Association in a resolution that "a national political party expressing the desires of the farmers and laborers be formed at once." It called upon leaders of progressive groups to meet at the earliest opportunity and promised support to any group that would promote and establish the Holiday program.[21]

Dr. Weiss's fateful bullet on September 8, 1935 removed forever the most magnetic of the political demagogues of the thirties. Long's assassination, declared Reno, "is only a sample of what we may expect to happen to all those who oppose the bureaucracy that has determined to overthrow representative government and the destruction of the rights and liberties guaranteed by the Constitution. . . ."[22]

The death of Huey Long made the task of unity no easier. Two major factions claiming to represent the left were pointing toward a third party movement in 1936; one, the Coughlin Social Justice movement, the other, the Farmer Labor Political Federation of Amlie, Williams, and Bingham. Milo Reno, despite his enthusiasm for Coughlin and Long, continued to work with the little Farmer-Labor party in Iowa established by Howard Y. Williams and insisted that they were basically the same movement. Both stood for cost of production, both condemned bankers, trusts, and what was considered the close alliance of the New Deal with corporate and financial interests. Yet here the congruence ended. One was emotional, irrational, given to name calling, evangelism, and panaceas; the other was an idealistic, humanitarian movement of those who believed the capitalist system must be replaced by a new social organization, both more efficient and more humane.

Milo Reno was an agent in the delivery of the third party movement into the hands of demagogues. Howard Y. Williams was one of the most broad-minded of men (he would have included Townsend supporters and Communists in the Farmer

[20] *New York Times,* April 28, 1935, p. 1; Robert Morss Lovett, "Hue [sic] Long Invades the Midwest," *New Republic,* LXXXIII (May 15, 1935), 10-12.

[21] *Farm Holiday News,* May 10, 1935.

[22] *Ibid.,* September 25, 1935.

Labor Political Federation), but after attending a Farmer-Labor meeting in Des Moines he reported sadly to Amlie, "It is very disappointing, however, to see the blind following of most of the leaders for Huey Long and Father Coughlin. . . . Milo Reno took the floor at the afternoon meeting and made a talk fully endorsing Coughlin and Long. Some of the delegates treated him rather roughly after the remarks but there is no doubt that he had strong support." [23]

Although differences remained more latent than patent, not all Farmers' Holiday members could follow in the direction Milo Reno was leading. The influence of John Bosch and the Minnesota Holiday, the largest of the remaining state units, was evident at the same April convention that listened to Long and endorsed a third party. For the first time, the platform of the Holiday Association called for a new economy based upon "production for use," the watchword of the Farmer-Labor party. The endorsement was a semantic one, for the resolution went on to define production for use as incorporating cost of production, cheap farm finance, and congressional control of the monetary system—the traditional slogans of the Farmers' Holiday.[24] For Reno this was probably only a necessary step in bringing about the amalgamation of protest groups, but the pages of the *Farm Holiday News* over the ensuing year reflected an increasing divergence of factions within the Association. A good illustration was the edition of December 26 which contained a passionate editorial by Reno indicting an administration that in his opinion had disregarded and subverted the Constitution to establish a government of men, not of laws. Side by side with Reno's editorial was a column by Richard Bosch of Minnesota condemning a literalistic interpretation and declaring "Like the Sabbath, the constitution was made for man, not man for the Constitution."

Perhaps inconsistencies could still be abided through 1935 in a vain hope that unity was still possible. John Bosch criticized the many small groups more interested in leadership than results and Thomas R. Amlie made a clear bid for Farmers' Holiday support when he wrote that while inflation would help farmers solve immediate problems, something more was needed: a replacement

[23] Williams to Amlie, April 2, 1935. Williams Papers.

[24] *Farm Holiday* News, May 10, 1935.

for the "wornout profit system."[25] Howard Williams persisted in 1936, over the protests of eastern leaders of the American Commonwealth Federation, in attempts to draw together Coughlin, Townsend, and Farmer-Labor factions.[26]

For his part, Milo Reno labored in Iowa to bring about what he considered a unity of progressive forces—the sum total of his efforts was to sell out the little Farmer-Labor party there to the Coughlin movement. L. M. Peet, business manager of the Iowa Farmers' Union and at the time probably Reno's closest friend and advisor, was chairman of the Social Justice Clubs of Iowa. A so-called unity meeting in February, 1936, included, in addition to the Social Justice Clubs, the Iowa president of the Townsend Clubs, Iowa Holiday delegates, the Farmer-Labor party, and representatives of the Twentieth Century Club (described as an association of Des Moines citizens of Italian descent). Milo Reno did not live to reap the harvest he had sown in the two places where his influence was greatest: the little Farmer-Labor party in Iowa and the Farmers' Holiday Association. Both split and crumbled over the Lemke candidacy in 1936.[27]

The third party movement did not operate in a political vacuum. Important changes on the right, the left, and in the center were radically altering the environment in which the protest movements operated. Both Father Coughlin and Milo Reno continued to appraise the New Deal in terms of the legislative program of the Hundred Days, yet by 1935 dramatic changes had rendered what had been principal emphases into little more than past history. The coup de grace administered to the NRA by the Supreme Court, the passage of a banking bill strengthening the government's role in the Federal Reserve System, the restriction upon business combinations in the Public Utility Holding Company Act, the new Frazier-Lemke Moratorium Act, and massive programs for relief and public works—all pointed to a new departure in New Deal legislation. The seeming government sympathy for restrictive business practices, exemplified by the NRA, had disappeared. The new focus was more upon welfare than planning

[25] *Ibid.*, August 10, September 10, 1935.

[26] Prospectus, Williams Papers, 1936.

[27] *Farm Holiday News*, March 10, 27, 1936; Minnie Duvall to Howard Y. Williams, September 6, 1936, Williams Papers.

and on regulation of business rather than cooperation with business. Most of these new measures served to correct some of the abuses that had most troubled dissident progressives.[28]

Collateral to the shift of the New Deal, forces of conservatism and reaction that had been slumbering during the early years of the administration reawakened. This was most manifest in the emergence of the American Liberty League. Regardless of the stir on the left, the most dangerous political threat to the New Deal was the Republican party which appeared in 1936 to be captive to some of the nation's most conservative interests. Progressives might well ask themselves, if they helped preside at the burial of the New Deal, what would be the alternative?

Problems with extremism of a different sort troubled progressives. With the new Popular Front policy, sincere and friendly Communists, warmly espousing their support for reform in place of revolution, were taking their places in organizations of the American left. The question of how to respond to them prompted serious questions and sometimes acrimonious discord that could polarize even a progressive organization. Moreover, partly because the Communists were present to point it out, the parallels of some of the strictures of American extremists to the anti-semitic, anti-democratic policies of European Fascists send a shudder through the ranks of American progressivism. Perhaps it really could happen here.

The Farmers' Holiday Association, an organization which had outlived its purpose, remained in 1935 only as a haven for die-hards and dissidents of many different persuasions. It was in its last days a staging ground in microcosm for the conflicting tendencies and purposes in American progressivism as the election of 1936 approached.

---

[28] This must be qualified by the fact there was no change in agricultural policy during 1935. The first real shift in legislative emphasis did not come until the Bankhead-Jones Farm Tenant Act of July, 1937 and the establishing of the Farm Security Administration in September of that year.

Thirteen

THE LAST DAYS

Reno was the voice of an attitude toward the farm people of our nation—are they to be merely the commercial producers of cheap food; or are they to be the self-respecting foundation of our civilization? Are they to be mere pawns (peasants) in our commercial structure? Or are they to be the representatives of the wholesome way of life on which shall be built physical and mental and moral soundness for a wholesome and enduring social order? [Wallace Short, *Farm Holiday News,* May 25, 1936].

A core of true believers remained steadfast to organizations of radical farm protest long after they ceased to have purpose and utility. These were not individuals attracted by rational economic motives. The very fact of their loyalty was proof these were men disaffected and unaccustomed to compromise. The ranks of those who could bear the stigma of supporting a futile cause or openly espouse communism in rural America must have included persons with deep ideological convictions and probably encompassed a fair share of cranks. Judging from names recurring in publications, those who remained to the end with Milo Reno were a hard-core of Iowa Farmers' Union members, some of whom had championed cost of production and agrarian radicalism even before 1932; the nucleus of the United Farmers' League and the latter-day Madison County Plan were likely to be individuals committed to radicalism, if not communism, since the days of the Non-Partisan League. When the Farmers' Holiday Association split in June, 1936, the issue was not principle against expediency; it was which brand of extremism to follow.

The history of the Farmers' Holiday Association after the disastrous autumn strike in 1933 dwells largely upon Milo Reno. For most of the adherents, even those in Minnesota who could not subscribe to all his tenets, this man symbolized the movement. And by reciprocation, the organization and Reno's personality were inseparably linked. Milo Reno's being and sustenance was in the folksy rural gathering, the friendship and adulation of farm people, the sleepless nights passing from one speaking engagement to another, the fervor of the crusade to save from destruction the small farm way of life. Could Milo Reno have borne loneliness or understood tragedy, his last years might have provided the raw materials out of which great dramas are shaped. For all his bold words, Reno knew the cause of farmer protest was dying. Friends report he was often seized by long moods of black depression; once a prohibitionist, he took to drinking heavily in the last years of his life and several of his radio broadcasts were canceled because he was incapacitated by intoxication. Listening to a recording of Reno's last speeches, one hears something different from the sonorous and firm timbre the words demand; one hears the waspish voice of a tired old man.[1] There is a suggestive parallel in the last days of two other evangelical rural leaders who both in role and ideology closely proximated the career of Reno. Note the sad spectacle of the defeated Bryan at Dayton, Tennessee or the near-insanity that haunted A. C. Townley to his death.

For Milo Reno to fight the good fight meant to stand unwavering for principle—in his instance, the abiding faith that inflation of the currency was the only sure route to economic well-being and the firm conviction that the government must guarantee to farmers cost of production prices. Milo Reno had the image of himself as a radical, and, in the sense that Populist individualism is radicalism, he was. He was consistently at war with whatever forces seemed to him to compromise the traditional individualism of the American farmer: for most of his career the enemy was conspiring middle men and Wall Street financiers who robbed the farmer of his economic independence; after 1933 the enemy was an expanding government bureaucracy that seemed in the same fashion to take away the farmers' individual initiative and enmesh them in forces beyond their control. Both Reno and his

[1] A recording of several of Reno's radio addresses is in the Reno collection, State University of Iowa.

Populist predecessors advocated programs that called for no small measure of government regulation and control, but neither of them ever advocated a single policy that compromised in any way the individual freedom of the farmer. Reno could not understand why his consistent stand for individualism carried him into common cause against the New Deal with the conservatives and business interests he had battled all of his life. Populism in the nineties stood for social changes that would enhance and protect the economic position of the family farmer. In the thirties, when technology, increasing consolidation, and the necessity for limitation of an overefficient production had rendered small farms an economic anachronism, those who defended Populist-style individualism fought a defensive action; they had to oppose social change. Indeed, the times had passed by one who had learned his political philosophy at the feet of James B. Weaver.

Fatigued and harassed by a persistent case of influenza, Reno entered a sanitarium at Excelsior Springs, Missouri, in March, 1936. "Tell them I'm really sick," he told his friend Dale Kramer as he departed. Kramer reflected that Milo Reno was a man who wanted to die.[2] In late April he talked with a reporter from the *Excelsior Springs Daily Standard,* the last interview of his life. One thing, he meditated, hounded and hounded him, ". . . the fact that the farmers, for whom he had given his entire life, would not cooperate to the extent that their problems might be solved." [3] Milo Reno died on May 5, 1936.

The legacy of Milo Reno was manifest in the two organizations to which he devoted the last sixteen years of his life, the Farmers' Holiday Association and the Iowa Farmers' Union. The former headed speedily to the inevitable rupture for which Reno had prepared the way and just two months after his death his Iowa followers bolted from the Association. The Iowa Farmers' Union by 1939 was a tiny schismatic group of 1,800, estranged from the National Union organization.[4]

---

[2] For factual data concerning Reno's personality, I am indebted to three friends of his: Dale Kramer, personal interview, March 22, 1962; John Bosch, personal interview, April 1, 1962; Frank Karstens, personal interview, August 23, 1961. Interpretations presented here, however, are my own.

[3] *Excelsior Springs* (Missouri) *Daily Standard,* May 5, 1936.

[4] Charles M. Wilpert to Page Hawthorne, January 8, 1940. Hawthorne Papers.

The breach in the Holiday Association was entangled in the fate of the Farmer Labor movement that had been building under the auspices of the Farmer Labor Political Federation (now the American Commonwealth Federation) since 1932. As the time approached when it was necessary to choose whether a third party movement would be launched in the 1936 election, three questions hobbled the decision: (1) Given the progressive orientation of the New Deal, should there be a third party at all? (2) Should Communists be permitted in the movement? (3) Was there any possibility of cooperation with the Coughlin political crusade? In answering these questions the leadership divided. Howard Williams and John Bosch, the Minnesota leaders, favored the convening of a national convention prior to the meetings of the two major parties. They sensed there was enough discontent with the New Deal in the West that such a meeting might induce one or both of the major parties to propose a radical platform of their own; a few midwestern leaders even felt a third party candidate might win the presidency. Leaders in the East were more pessimistic; organized labor under the leadership of John L. Lewis and Sidney Hillman was being primed to throw its support to Roosevelt; they discounted much of the radical activity in the West because of the seemingly irreconcilable factionalism of the dissenting groups. Moreover, it appeared that two key midwestern leaders, Floyd B. Olson and Robert LaFollette, Jr., were planning to support Roosevelt.[5]

Interested in some form of political offensive, Williams and Bosch in their concern for unity were willing to include Communists, now vigorously urging a Popular Front Farmer-Labor party, in the projected conference. Alfred Bingham and Representative Tom Amlie were adamantly opposed. "We who ourselves have been victims of red-baiting should not become red-baiters," Williams wrote. "At our state convention of the Farmer Labor Party two weeks ago a number of Communists sat as delegates and made a real contribution." [6]

Disagreements came to a showdown when Minnesota Farmer-

---

[5] McCoy, 101-7; these themes occur in many of the letters to Howard Y. Williams, January-May, 1936.

[6] Nathan Fine to Howard Williams, March 4, 1936; Williams to Fine, March 22, 1936; Williams to Mrs. George F. Brown, April 15, 1936. Williams Papers.

Labor leaders, unwilling to await the hesitant eastern liberals, called a conference to meet in Chicago, May 30-31. Williams, who was largely responsible, disavowed any notion that the meeting was for the purpose of launching a presidential ticket; it would focus on strengthening state groups and attempt to galvanize progressive sentiment as a way of placing pressure on the major parties.[7] The eastern leaders were irate. One complained not only had he not been notified but he had first read of the conference in the *Daily Worker*. Floyd B. Olson, who had signed the conference call on the understanding it would be concerned only with state tickets, withdrew in late May, declaring that further consideration of a third party presidential ticket might defeat a liberal administration and elect "Fascist Republicans." [8]

As the date of the conference approached, the central bone of contention was Communist participation. David Dubinsky, Paul Douglas, Alfred Bingham, and Thomas R. Amlie boycotted the meeting. While the conference was not, as some of the seceders insisted, dominated by Communists, it dovetailed with Communist efforts to end their isolation and join in a broad Popular Front party. The *Daily Worker* hailed the meeting and Earl Browder made an appearance to pledge party support to any third party effort.[9] The platform adopted called for opposition to monopolies, social security, public ownership of natural resources, a thirty-hour work week, a curb on the powers of the Supreme Court, the Frazier-Lemke bill, and doing away with further crop reduction. Whatever the objectives of the conference may have been, it was a stillborn effort. Of the old line leaders in the American Commonwealth Federation, only Howard Williams and John Bosch attended. A motion calling for a third party convention in the near future was amended to leave the decision to the Minnesota Farmer-Labor party and the advisory council, for all practical purposes killing the proposal.[10]

To crown the frustrations of midwestern progressives, on June 19, William Lemke, hero of the martyred Frazier-Lemke bill, announced his candidacy for the presidency on the Union Party ticket, and Father Coughlin, the instigator, issued a public en-

---

[7] Williams to Amlie, May 7, 1936. Williams Papers.

[8] McCoy, 107; Schlesinger, *Politics of Upheaval*, 549.

[9] *Daily Worker*, June 3, May 14, 1936.

[10] McCoy, 111-13.

dorsement.[11] Usher Burdick, veteran of the Farmers' Holiday, became his campaign manager and E. E. Kennedy, national secretary, announced that the Farmers' Union would support Lemke.[12] Farmer-Labor leaders were now in hopeless confusion. Lemke's candidacy promised to siphon off much of the agrarian support upon which they had counted. Moreover, many of them, including Williams and Bosch, grew suspicious that the Coughlin program contained fascist ingredients. Faced with the uninviting prospect of a movement that could command little support, whose only accomplishment would be to draw progressive votes away from Roosevelt, the Farmer-Laborites chose the only remaining alternative. One by one, Olson, Amlie, Bingham, Fine, Bosch, and Williams endorsed Roosevelt.[13]

With the National Farmers' Union committed to the Lemke candidacy, with Communists intermingled in the membership of its largest state constituency, and with the Farmer-Labor movement fast aligning itself with Roosevelt, the Farmers' Holiday Association assembled for what would prove to be its final national convention. John Bosch, who had succeeded Reno as president, justified the change in convention locale from Des Moines to Minneapolis on the grounds the organization should meet in the state with the largest membership. The convention call, announcing the purpose would be to "seek to unite all farmers who believe the profit system has outlived its usefulness," indicated a different orientation from the assemblies of past years that in their turn had acclaimed first Father Coughlin and then Huey Long.[14] Two issues confronted the 300 delegates. For the Minnesota group, the most important objective was to prevent the organization from going the way Milo Reno would have desired—to line up with the Farmers' Union behind Lemke. To the little band of Reno followers, mostly from Iowa, the vital thing was to save their Holiday Association from Communist control. In short, for the die-hard adherents who remained loyal, ordinary political alternatives had little meaning; the issue was which brand of extremism would prevail.

---

[11] Schlesinger, *Politics of Upheaval,* 555; Blackorby, *Mississippi Valley Historical Review,* 76.

[12] *Daily Worker,* June 25, 1936.

[13] Schlesinger, *Politics of Upheaval,* 596.

[14] *Farm Holiday News,* May 25, June 27, 1936.

Hardly had the delegates settled in their seats when a bitter floor fight began over the method of voting. John Chalmers, organizer of the first Iowa Holiday meeting four years earlier and a Lemke backer, insisted that consistent with past convention practice each state delegation, regardless of the number in it, should have one vote. The leaders of the Minnesota delegation, far the largest attending, insisted that everyone present should vote, the majority ruling. No copy of the by-laws was available; a check would have shown that Chalmers was correct. The debate over voting, however, was a front for the major issue. Voting by delegation would have given the New Mexico president, whose association numbered eighteen members, one vote; the Maryland organization, represented by its only member, would likewise have cast a vote. Since these phantom presidents were Lemke supporters, to have so voted would have meant an endorsement of the Union party. On the other hand, mingling in the Minnesota and North Dakota delegations were men who until a few months ago had been the most active leaders of the United Farmers' League: to vote by head would enfranchise Communists and fuse them into the Holiday Association. The resolution of the issue rested with the presiding officer, John Bosch. The *Farmers' National Weekly* described the situation with something less than impartiality:

> . . . President Bosch got out of his chair and came in front of the table on the platform, and while the crowd sat hushed, thinking that the democracy of the convention stood in the balance, Bosch said: 'If the affairs of the Holiday Association cannot be trusted to the people who are its members, I don't want to be its chairman.' It was there. It had been said, and the house went wild with applause. . . . For once in the history of the militant farm organizations of America—the President stood out for rank and file control.[15]

[15] July 10, 1935. The sharp divergence of these two factions in the last days of the Farmers' Holiday Association suggests a hypothesis that might be applicable to earlier agrarian movements, particularly Populism. Richard Hofstadter, *The Age of Reform* (New York, 1955), emphasizes that "the utopia of the Populists was in the past, not the future" (p. 62). Norman Pollack, *The Populist Response to Industrial America* (Cambridge, 1962), 1-12, stresses, "Populism formulated an extraordinarily penetrating critique of industrial society" (p. 9). The obvious point is that both these tendencies were present. A neglected facet of the history of farmer movements may be the recurring conflict between factions within the movement demanding immediate solutions and those pursuing long-range social reform goals. The schism that divided and destroyed the Farmers' Holiday Association seems

It *was* there. With the Lemke candidacy spurned and known Communists voting, the Farmers' Holiday Association was no longer a hospitable place for followers of Milo Reno. Five angry state chairmen, those from Iowa, North Dakota, New Mexico, Wisconsin, and Maryland, seized the records, signed a querulous statement condemning the sellout of the organization to "Communists" and withdrew. According to press reports, the leader of the bolters was Usher Burdick, Lemke's campaign manager, but this was in error. For reasons unknown, Burdick did not join the dissidents. A few weeks later he urged the North Dakota convention to support the national and himself moved and won unanimous consent for a resolution of confidence in John Bosch.[16]

The dissidents gone, the convention hastened to adopt a platform endorsing the principles of the Farmer-Labor party, slightly pinkened by a Popular Front color. For the first time in the history of the Holiday Association the words "cost of production" were lacking. The new platform called for the Frazier-Lemke bill, reassertion of the power of Congress over the issuance of money, production for use, and a graduated land tax. A new plank, never present while Milo Reno lived, urged the organization of tenant farmers. The Association joined the *Daily Worker* and other Popular Front groups in a campaign against William Randolph Hearst and it joined that most popular of Popular Front associations, the American League Against War and Fascism.[17]

The convention completed the process of amalgamation of the Communist agrarian movement into the Farmers' Holiday Association. In the last that was heard of the Madison County Plan, its president joined it to the national Association. The United Farmers' League disappeared. On August 28, the *Farm Holiday News* and *Farmers' National Weekly* fused into a new publication, the *National Farm Holiday News*. Headquarters of the paper was

---

to reflect a conflict of purpose that has occurred in all agrarian movements. The conflict in the movement of the thirties is sharpened by the link of both factions with external movements, i.e., the Coughlin movement and the Communist party.

[16] *Des Moines Register*, July 3, 1936; *Farm Holiday News*, July 25, 1936. Twenty years later when Mrs. Helen Wood Birney revived the memory of the Farmers' Holiday Association by charging it had been Communist dominated, Usher Burdick, at the time representative-at-large from North Dakota, vehemently denied the charges. *Des Moines Register*, March 25, 1954.

[17] *Farm Holiday News*, July 11, 1936.

Minneapolis, where the *Weekly* had been published and the new publication followed the format of the old FNCA paper. A grant from the Garland Fund, a source of support for various left wing enterprises, supplied 30 per cent of the operating cost of the new publication. Two weeks after the convention, Bosch told the North Dakota Holiday meeting that only 10 per cent of those at the Minneapolis meeting were Communist affiliates.[18] He underestimated. Of twenty-seven individuals appointed to committees at the Minneapolis convention, eight of them had been associated with the Madison County Plan or United Farmers' League. If this sample is representative, the proportion was closer to one-third.[19] In one sense, the union of the organizations was a victory for the Communists. By a policy of patience and waiting they had gained at this late date a respectable position in the farmers' protest movement. But in a larger sense this was no victory at all. The farmers' protest barely existed. A share in a schismatic little organization with little functional purpose was sparse remuneration for three years of hard work, ideological vicissitudes, and frustration.[20]

Was this latter-day Farmers' Holiday a Communist front? No party agents stood behind the leaders to direct and guide them. John Bosch controlled the organization and acted as its spokesman; Dale Kramer continued as editor of the new *National Farm Holiday News*. Lem Harris wrote occasionally for the paper and, although he was recruiting members for the party at the time, he did this outside of the Holiday Association. Communists did not need to control the Holiday. Its stands for farmer unity, opposition

[18] *Farmers' National Weekly,* August 21, 1936; *National Farm Holiday News* (Minneapolis), August 28, 1936; Lem Harris, personal interview, April 11, 1962.

[19] *Farmers' National Weekly,* July 10, 1936 lists the committee appointments upon which I base this conclusion.

[20] A parallel in Communist strategy toward labor organizations may be noted. Glazer, 214-15, observes: "Each time the membership of the various unions [organized in the first post–New Deal upsurge of labor organizations in 1933 and 1934] dropped . . . the radicals of all shades and varieties hung on and consolidated their strength by taking over key posts. While the ordinary worker would join or leave a union in response to economic conditions, to the chances of the union to improve his state, and so on—in effect, for rational economic reasons—the radicals would join for reasons of ideological commitment. What few there were, were generally to be found in the unions, and consequently their weight in the union was obviously greater than their weight in the industry."

to fascism, and demands for more generous relief payments were in perfect accord with the moderate radicalism and spirit of cooperation that marked the Popular Front.[21]

The editorial policy of the *News* in the autumn of 1936 reflected the confusion and dismay of the shattered Farmer-Labor progressives. The pre-election editorial gave this ringing endorsement: "Of the two candidates with any chance of being elected, one is only slightly more reactionary than the other, the main difference being that the worst of the fascist-inclined reactionaries have lined up with Landon." [22] There was no equivocation, however, toward Lemke and Coughlin. No acknowledged progressive, including Senator Frazier or the late Governor Floyd B. Olson,[23] an editorial declared, supported Lemke. He would aid Landon by drawing away progressive votes. His program had a liberal-reactionary vagueness that smacked of fascism. Noting the presence among Lemke backers of Newton Jenkins, an acknowledged Chicago Fascist, and former mayor "Big Bill" Thompson (Union party gubernatorial candidate in Illinois), the editorial concluded, "If Coughlin and Lemke are not Fascists, fascists certainly are largely in control." [24] In a widely publicized interview, Dale Kramer talked with Father Coughlin. The priest told him, "We are at the crossroads. One road heads toward fascism, the other toward Communism. I take the road toward fascism." Were a Communist government to be elected, Coughlin declared, he would be "out fighting it with a gun." [25]

Old Reno followers waged a campaign of invective against the new Holiday Association that would have done justice to the master himself. The bolters' main strength was in Iowa; North Dakota, at Burdick's behest, remained in the Association and the Wisconsin Holiday disintegrated in a bitter convention battle over the Lemke candidacy.[26] The *National Farm Holiday News*,

---

[21] John Bosch, personal interview, April 1, 1962; Dale Kramer, personal interview, March 22, 1962; Lem Harris, personal interview, April 11, 1962.

[22] October 30, 1936.

[23] Governor Olson died in August, 1936. The editorial was mistaken regarding Senator Frazier, who was a Lemke supporter. Blackorby, *Mississippi Valley Historical Review*, 77.

[24] September 25, 1936.

[25] *National Farm Holiday News*, September 25, 1936; Schlesinger, *Politics of Upheaval*, 629; Dale Kramer, personal interview, March 22, 1962.

[26] *National Farm Holiday News*, September 25, 1936.

notably broad-minded, printed some of the most scurrilous attacks of the estranged group. "A paper through which Milo Reno carried on his great fight for cost of production and for the Frazier-Lemke Bill seems to have joined hands with the banker controlled Farm Bureau and have become just an ordinary Jim Farley rag," wrote a Wisconsin correspondent.[27] E. E. Kennedy, an old Reno ally and architect of the Farmers' Union endorsement of Lemke, made the accusation that "since the Holiday's great leader, Milo Reno, passed away . . . a group maintaining the name and form of the . . . Association have, to all intents and purposes, abandoned and discarded the principles and program to which the association had steadfastly adhered since its inception. . . . In May, 1936 . . . the leadership . . . was quickly seized by a non-farmer element within the organization with strong Communistic tendencies."[28] John Bosch retorted that Kennedy had led the Farmers' Union off the progressive path and "placed it on a road which might very possibly lead to fascism." Perhaps this criticism had some effect for the National Farmers' Union, humiliated by the disastrous failure of its one venture into third party politics, replaced Kennedy as secretary a few weeks later.[29] Dale Kramer drew the unenviable assignment of attending the Iowa State Farmers' Holiday convention on October 30. Twenty-five years later he still retained vivid memories of the meeting—he considered himself fortunate to have escaped lynching. Over Kramer's protests the seventy-five irreconcilable Reno supporters who attended refused to recognize the elections at Minneapolis and disavowed the *National Farm Holiday News.* John Chalmers called Kramer "un-American" and "in the pay of the present administration."[30]

In its last breath of political activity, the Iowa Holiday contributed to the imbroglio of third party politics in the state in 1936. Ninety-five per cent of the delegates at the state convention

---

[27] October 2, 1936.

[28] *National Farm Holiday News,* November 12, 1936.

[29] *Ibid.; Farmers' Union Herald,* December, 1936. Kennedy operated a legislative news service for a short time in Washington, D.C. In 1942 he was director of research for the United Dairy Farmers Division of the United Mine Workers. McCune, 196-97. In the 1950's he was an official in the Washington headquarters of the AFL-CIO.

[30] *National Farm Holiday News,* November 13, 1936; Dale Kramer, personal interview, March 22, 1962.

of the Farmer-Labor party were Holiday and Union members; the party, founded originally with the help of Howard Y. Williams, endorsed Lemke. It did not succeed, however, in coalescing with the Union party and in several congressional districts, both third parties entered candidates. Chairman of the Union Party in Iowa was L. M. Peet of the Iowa Farmers' Union; A. J. Johnson, president of the Union, and W. C. Daniel, once president of the Woodbury County Farmers' Holiday, were Union party congressional candidates. John Chalmers was paid a salary of $200 weekly by Harrison Spangler, Republican National Committeeman, to campaign in Iowa—for Lemke. All this was to little avail. Lemke polled only 2.6 per cent of the Iowa vote; even then his strength was largely in the Catholic urban centers of eastern Iowa rather than in farming counties.[31]

No written record or even clear recollection recalls the fate of the Iowa Farmers' Holiday Association; it died quietly with the ill-fated Lemke candidacy. The heritage of Milo Reno lived on in the Iowa Farmers' Union. By 1939, it consisted of only 1,800 members and was still led by men who had fought with Milo Reno for cost of production.[32] The National Farmers' Union returned in 1937 to the control of the cooperative faction beaten by Simpson and Reno in 1931; political battles were forgotten and the organization made its peace with the AAA. The Iowa unit was a tiny and estranged state body.[33] The editorials in the *Iowa Union Farmer* in 1939 suggest where Milo Reno's policies might have led. One month before war began in Europe the editor wrote that Hitler had utilized the nation's resources to put people back to work. He explained that while he did not favor the Nazi dictator, he could not refrain from praise for his accomplishments. A few months later, another editorial declared American newspapers were controlled by Wall Street financiers and Jews who

[31] Minnie Duvall to Howard Williams, September 6, 1936; A. J. Johnson, personal interview, March 12, 1962; W. C. Daniel, personal interview, March 14, 1962; John Chalmers, personal interview, October 21, 1961. Wallace Short of Sioux City was the Farmer-Labor candidate for governor. See Robinson, *They Voted for Roosevelt*, 89-93.

[32] Iowa Farmers' Union, Minutes of Councillor's meeting, June 29, 1937, March 29, 1939, Hawthorne Papers. For the subsequent history of the Iowa Farmers' Union see Steven A. Chambers, "Relations Between Leaders of the Iowa and National Farmers Union Organizations, 1941 to 1950" (unpublished honors paper, Department of History, State University of Iowa, 1961).

[33] McCune, 200-214; *Farmers' Union Herald*, December, 1937.

in league with Jews and Communists in Hollywood were conspiring to discredit the Nazi regime in Germany.[34] Thus perished the last vestige of Milo Reno.

The Farmers' Holiday Association with headquarters in Minneapolis continued for a year after the election of 1936. A fringe group even before the split, merger with the Farmers' National Committee for Action did little to offset membership losses. The Minnesota Holiday had 1,876 dues-payers in 1936. No further statistics were ever released, but the largest single county, Lac Qui Parle, which had 536 members listed in the 1936 report, had only 150 in July, 1937. Conventions were held in North and South Dakota in 1937 and the former organization seemed to revive, at least in publicity output, in the spring and summer.[35]

Consonant with its Popular Front character, the social program of the Holiday Association shifted leftward. The problems of tenant farmers, largely ignored for four years, was a major new concern. Fifty per cent of American farmers are tenants, Bosch told the Minnesota Farmer Labor Federation; any new agricultural program must include them. To this end, a model farm tenancy bill was proposed which would permit tenants to purchase farms by payment in kind to landlords. Oliver Rosenberg, new president of the North Dakota Holiday, defended the plan in congressional hearings in January.[36]

The most interesting of the final ventures of the Holiday Association was an attempt to consolidate radical farm organizations and hopefully join them in a united front with the growing labor movement. (The objective was suggestive of the united proletarian front the Farmers' National Committee for Action had attempted three years earlier.) John Bosch called for a merger of the Holiday, the Farmers' Union, the Share-Croppers' Union (a Communist organization), and the Southern Tenant Farmers' Union (a Socialist organization).[37] The Holiday joined in a cam-

---

[34] October 27, November 7, 1939.

[35] National Farmers' Holiday Association, Minnesota Division, Report of the Annual Convention, October 29-30, 1936, FNCA files; *National Farm Holiday News,* July 16, 1937. Dale Kramer, personal interview, March 22, 1962.

[36] *National Farm Holiday News,* January 13, February 26, 1937. Oliver Rosenberg, was the brother of Anton Rosenberg of Newman Grove, Nebraska, one of the organizers in 1932 of the Madison County Plan.

[37] *National Farm Holiday News,* December 11, 1936.

paign which, if successful, would have marked a milestone in American agrarian history. Some CIO organizers, during the period of expansion in 1937, broached the idea of incorporating farmers into their growing union. "The American workers and the American farmer have a common goal," wrote John L. Lewis [38] and an unnamed leader declared, "I am of the opinion that the Holiday Association may become the CIO of the farmers." [39] The idea moved beyond the planning stage. John L. Lewis, John Bosch, and James Patton, secretary of the Colorado Farmers' Union (soon to become president of the National) attended the organizing convention of the United Cannery, Agricultural, Packing and Allied Workers' Union in July, 1937. The active instigators of the conference were some of the small legion of Communists who in the early days of the CIO occupied crucial positions as organizers. John Brophy, national director of the CIO, one of the leaders at the meeting, was an ex-radical and while no Communist himself, he was responsible for bringing many of the party members into CIO work. The president of the Agricultural Workers' Union was Donald Henderson, a party member. John Bosch, who was having some misgivings about the company he was keeping, was not optimistic: "So the plea I would make to your convention," he told the delegates, "above all other things is, do not forget that this middlewest . . . has supported the legislation you have demanded . . . if you alienate that part of the United States, I am absolutely firmly convinced that the farmer who could become your ally will become the backbone of fascism in the United States." [40] The Agricultural Workers' Union made some attempts to organize farmers and for a few years there was some liaison between the CIO and the National Farmers' Union, but that is beyond the scope of this study. The Farmers' Holiday Association was a weak bridge over which to carry unionism to family farmers. The CIO never crossed it.

The new face of the Farmers' Holiday attracted little enthusiasm among midwestern farmers. Economic inequities certainly re-

[38] *The C.I.O. and the Farmers,* extracts from speeches made at the First National Convention of the United Cannery, Agricultural, Packing and Allied Workers' Union of America, Denver, Colorado, July 10, 1937 (pamphlet in possession of the writer).

[39] *National Farm Holiday News,* July 30, 1937.

[40] *The C.I.O. and the Farmers, loc. cit.;* Howe and Coser, 372, 375.

mained in 1937, but nothing short of compelling need could have driven farmers into anything so counter to usual rural tendencies as an alliance with organized labor. President John Bosch grew disillusioned. After attending a conference of agricultural leaders in Washington, he returned to announce that the AAA was willing to go further than most farm leaders. He announced his support of the ever normal granary proposed by Henry Wallace. In contrast to the call to convention a year earlier of "all farmers who believe that the profit system has lost its usefulness," Bosch now averred, "The Farmers' Holiday Association realizes that we still exist within the capitalist system, and that it is necessary for farmers to try to protect themselves against the exploitation of that system. It is necessary to maintain a control of the flow of commodities to market, if one is to maintain a price on those commodities." [41] Adding to Bosch's discomfiture were the circumstances surrounding the annual "People's Lobby" at St. Paul in early April. Bosch organized the march at the request of Governor Elmer Benson. Some undisciplined participants jammed the Senate chamber, discarded sandwich wrappers on the floor, and refused to move. The press, hostile to Benson in the first place, emphasized the unpleasant details. Bosch, in embarrassment, presented himself for arrest. The offer was refused, but he felt he had been left "holding the bag" by Benson and his political supporters.[42] Speaking in July to the League for Industrial Democracy in New York, Bosch expressed the fear that the CIO was organizing too rapidly. "It is ten times easier to organize farmers against than for labor," he complained. "Unless we dramatize our cause, I fear very much that we won't have a farmer-labor party." [43]

[41] *National Farm Holiday News,* February 19, 1937. Bosch was correct that the administration was ahead of the farm organization leaders. The President's Farm Tenancy Committee, appointed in November, 1936, had taken the initiative in meeting a problem ignored not only in previous agricultural policy but by all major farm organizations. Out of this study emerged the Bankhead-Jones Act of June, 1937 and the expansion of the earlier Resettlement Administration into the Farm Security Administration. By its new attention to the plight of those among the nation's farmers who were "ill-fed, ill-clothed and ill-housed," the administration was not only undercutting the support, but destroying the necessity for radical farmer organizations.

[42] *National Farm Holiday News,* April 9, 1937; John Bosch, personal interview, April 1, 1962.

[43] *National Farm Holiday News,* July 16, 1937.

The hourglass was running dry. In Minnesota, three organizations, the two schismatic wings of the Farmers' Union plus the Holiday Association, competed for the allegiance of the progressive farmer. In November the long acrimony ended and the three merged to create one state unit of the National Farmers' Union. The parent organization in its convention at Oklahoma City the same month approved a resolution (with only one old Reno supporter dissenting) inviting the Holiday Association back into the fold.[44] One afternoon in late November a few intrepid farmers in Brown County, Minnesota gathered in a farmyard and succeeded in stopping a foreclosure sale; theirs was the last recorded action in the history of the Farmers' Holiday Association.[45] The newspaper published its last edition on December 31, 1937. Thus, the Farmers' Holiday Association, sometimes wayward child of violence and disorder, its visage much altered in five years of turbulent life, disappeared back into the Farmers' Union from which it had sprung.

Within the Union, the memory of Milo Reno and John Simpson was as forgotten as free silver at sixteen-to-one or cost of production. The problem of preserving the family farm way of life had entered an entirely new dimension. Benefit payments, acreage allotments, and price supports were a necessary and accepted method of operation and the choice was no longer between different alternatives, but between methods of administering one already chosen. The Union, although still more than any other agricultural organization the voice of the small farmer, was a firm supporter of the AAA. During the war years it gained new stature and respectability in Washington by supporting wartime price control measures over the opposition of the Farm Bureau Federation. Political positions of the thirties were reversed: as the Farm Bureau grew progressively disenchanted with price support measures, the Union's enthusiasm grew.[46]

Milo Reno was the last ideological radical to fight the farmers' battle with the old Populist principles of economic individualism and unqualified opposition to elitism in the form either of monopoly or bureaucracy. Looking to the past, not to the future, he

[44] *Ibid.*, November 19, December 17, 1937.

[45] *Ibid.*, December 3, 1937.

[46] McCune, 206-7; Christiana McFadyen Campbell, 169-70, 186. For example, the Farmers' Union supported the Farm Security Administration; the Farm Bureau opposed it.

was defending not just an economic system, but a way of life. Reno spoke for a day that had been; the radicals of the Popular Front spoke for a day that never came. The New Deal was the catalytic agent of time and change that defeated both of them. The agricultural program removed the rural discontent which provided Reno's ideology with the only driving force that could sustain it; benefit payments and price supports rendered the steady force of social and technological change that absorbs and destroys the family farmer far more painless than the radical left could have imagined possible.

BIBLIOGRAPHY

Manuscript Collections

Day, Vince Papers. Minnesota State Historical Society, St. Paul. The most useful parts of this collection were the frequent memoranda to Governor Floyd B. Olson, for whom Day served as private secretary.

Farmers' National Committee for Action files. Private collection of Mr. Lem Harris, New York City. This valuable collection consists of letters, minutes, pamphlets, and miscellany pertaining to the Communist party farm organization efforts in the nineteen thirties. Mr. Harris has agreed to eventually deposit the collection at the Library of the State University of Iowa.

Hawthorne, Page Papers. Library of the State University of Iowa, Iowa City. Hawthorne was president of the Iowa Farmers' Union in 1939-40.

Herring, Clyde Papers. Library of the State University of Iowa, Iowa City. This is a small, selected collection containing one valuable file on the declaring of martial law in northwest Iowa counties in 1933.

Kriege, J. Fred Papers. Private collection of Mr. John Kriege, Hayward, California. Mr. Kriege has agreed eventually to deposit the papers of his father, one of the organizers of the Farmers' Holiday in Nebraska, with the Nebraska State Historical Society, Lincoln.

Ormsby, George collection of Farmers' Holiday papers and clippings. Private collection, Wilton Junction, Iowa. Mr. Ormsby has made a lifetime avocation of gathering materials on farmers' movements and organizations. His valuable collection has been promised to the Library of the State University of Iowa, Iowa City.

Reno, Milo collection. Library of the State University of Iowa, Iowa City. This collection consists of scrapbooks and miscellany presented to the library by Reno's family.

Reno, Milo Papers. Library of the State University of Iowa, Iowa City. This collection contains all correspondence to Reno and carbon copies of his replies for the year 1933, plus miscellaneous scripts for editorials, articles, and speeches. Originally the property of the U.S.

Farmers' Association, Des Moines, Mr. Fred Stover, the president, and the Board of Directors agreed to its deposit at the library.

Roosevelt, Franklin D. Papers. Franklin D. Roosevelt Memorial Library, Hyde Park, New York. This study made use of the President's Personal File, the Official Files (Agriculture) and miscellaneous correspondence with individuals.

Williams, Howard Y. Papers. Minnesota State Historical Society, St. Paul. This is a valuable and well-organized collection of materials pertaining to the League for Independent Political Action and the American Commonwealth Federation.

Personal Interviews

Anderson, Mr. and Mrs. Clarence, son-in-law and daughter of Andrew Dahlsten, organizer of the Madison County Plan Farmers' Holiday Association. March 8, 1962.

Barlow, Lester, founder, The Modern Seventy-Sixers. January 4, 1962.

Bosch, John, president, Minnesota Farmers' Holiday Association; president, National Farmers' Holiday Association. April 1, 1962.

Chalmers, John, president, Iowa Farmers' Holiday Association. October 21, 1961. (Mr. Chalmers died November 26, 1961.)

Daniel, W. C., president, Woodbury County Farmers' Holiday Association. March 14, 1962.

Daugaard, Anton, member, Farmers' Holiday Association, Monona County, Iowa; arrested for Holiday activity in May, 1933. March 11, 1962.

Gustavson, Joe, participant in Farmers' Holiday picketing at Des Moines, August, 1932. August 1, 1961.

Harris, Lem, executive secretary, Farmers' National Committee for Action. April 11, 1962; July 15, 1963.

Haugland, Harry, chairman, Tri-County Council of Defense, Chippewa, Lac Qui Parle, and Yellow Medicine Counties, Minnesota. April 5, 1962.

Johnson, Andrew J., president, Iowa Farmers' Union; arrested for Holiday activity in May, 1933. March 11, 1962.

Karstens, Frank, member, Iowa Farmers' Union. August 23, 1961.

Kramer, Dale, editor, *Farm Holiday News*. March 22, 1962.

Lux, Harry, organizer, Madison County Plan Farmers' Holiday Association. March 1-2, 1962. (A duplicate of this tape-recorded interview is held by the Nebraska State Historical Society, Lincoln. Mr. Lux died October 13, 1962.)

Murphy, Donald R., editor, *Wallace's Farmer and Iowa Homestead*. March 15, 1962.

O'Connor, Edward L., attorney general of Iowa in 1933. March 21, 1962.

Ormsby, George, participant in the Cedar County Cow War. August 23, 1961.

Reck, I. W., president, Sioux City Milk Producers Cooperative Association. March 12, 1962.

Stover, Fred, president (1965), U.S. Farmers' Association, Des Moines. July 31, 1961; March 25, 1962.

Witt, Jack, member, United Farmers' League, Wisconsin. March 24, 1962.

Unpublished Material

Chambers, Steven A. "Relations Between Leaders of the Iowa and National Farmers Union Organizations, 1941 to 1950." Unpublished honors paper, Department of History, State University of Iowa, 1961.

Dahlsten, Andrew. "Holiday Association of Nebraska Organized Under the Madison County Plan: Origin, Purpose, Plan of Organization and Method of Procedure." Unpublished MS, n.d. (copy in the writer's possession).

Filley, Horace Clyde. "Effect of Inflation and Deflation upon Nebraska Agriculture, 1914 to 1932." Unpublished Ph.D. dissertation, University of Minnesota, 1934.

Korgan, Julius. "Farmers Picket the Depression." Unpublished Ph.D. dissertation, American University, 1961.

Lawrence, Howard Wallace. "The Farmers' Holiday Association in Iowa, 1932-33." Unpublished M.A. thesis, State University of Iowa, 1952.

Schonbach, Morris. "Native Fascism During the 1930's and 1940's: A Study of Its Roots, Its Growth, and Its Decline." Unpublished Ph.D. dissertation, University of California at Los Angeles, 1958.

Letters to the Writer

Barlow, Lester. October 19, 1961.

Lux, Harry. May 4, 1962.

Sorenson, Holger. January 29, 1962.

Turner, Governor Dan W. October 15, 1961.

Wallace, Henry A. November 28, 1961.

Newspapers

The Daily Worker (New York).

Des Moines Register.

Des Moines Tribune.

Excelsior Springs (Missouri) *Daily Standard.*

Farm Holiday News (St. Paul, February, 1933–January, 1934; Marissa, Illinois, February, 1934–December, 1935; Ames, Iowa, December, 1935–August, 1936).

Farm News Letter (Washington, D.C.).

Farmers' National Weekly (Chicago, January 30–November 10, 1933; Minneapolis, January 15, 1934–August 21, 1936).

Farmers' Union Herald (St. Paul).

Grand Island (Nebraska) *Daily Independent.*
Indianapolis (Indiana) *Star.*
Iowa Union Farmer (Columbus Junction, Iowa).
Lemars (Iowa) *Globe-Post.*
Lemars Semi-Weekly Sentinel.
Lincoln (Nebraska) *Herald.*
Lincoln Star.
Minneapolis Journal.
National Farm Holiday News (Minneapolis).
National Farm News (Washington, D.C.).
Nebraska State Journal (Lincoln).
Nebraska Union Farmer (Omaha).
The New York Times.
New York Evening Journal.
Norfolk (Nebraska) *Daily News.*
Norfolk Press.
O'Brien County Bell (Primghar, Iowa).
Omaha World-Herald.
Producers' News (Plentywood, Montana).
Sioux City (Iowa) *Journal.*
Sioux City Tribune.
Unionist and Public Forum (Sioux City).
Wallace's Farmer and Iowa Homestead (Des Moines).
Washington (D.C.) *Daily News.*
Willmar (Minnesota) *Tribune.*
Yale Daily News (New Haven).

Public Documents

Home Building and Loan Association v. Blaisdell, 290 U.S. 251 (1934).

Lux v. Nebraska, 126 Nebraska 133 (1934).

Nelson v. Doll, 124 Nebraska 523 (1933).

State of Iowa. *Biennial Report of the Adjutant General for the Years 1933 and 1934.* Des Moines: 1934.

State of Nebraska. *House Journal.*

U.S. Bureau of the Census. *Fifteenth Census of the United States: 1930. Population*, Vol. VI. Washington: Government Printing Office, 1933.

U.S. *Congressional Record.*

U.S. Department of Agriculture, Weather Bureau. *Climatological Data, Iowa Section.* Vols. XLII (1931) and XLIII (1932).

U.S. Department of Agriculture. *Yearbook of Agriculture, 1928.* Washington: Government Printing Office, 1928.

U.S. Department of Agriculture. *Yearbook of Agriculture, 1932.* Washington: Government Printing Office, 1932.

U.S. Department of Agriculture. *Yearbook of Agriculture, 1935.* Washington: Government Printing Office, 1935.

U.S. Department of Agriculture. *The Farm Real Estate Situation, 1933-34.* Circular 354. Washington: Government Printing Office, 1935.

U.S. House of Representatives, Committee on Un-American Activities. *Investigation of Communist Activities in the Chicago Area.* 3 parts. 83rd Cong., 2nd Sess., 1954.

U.S. Senate, Committee on Agriculture and Forestry. *Agricultural Emergency Act to Increase Farm Purchasing Power.* Hearings on H.R. 3835. 73rd Cong., 1st Sess., 1933.

U.S. Senate, Committee on Agriculture and Forestry. *To Abolish the Federal Farm Board and Secure to the Farmer Cost of Production.* Hearings on S. 3133. 72nd Cong., 1st Sess., 1932.

U.S. Senate, Subcommittee of the Committee on Agriculture and Forestry. *To Establish an Efficient Agricultural Credit System.* Hearings on S. 1197. 72nd Cong., 1st Sess., 1932.

Bulletins and Pamphlets

Black, A. G.; Shepherd, Geoffrey; Schultz, Theodore W.; Murray, William G.; Bentley, Ronald C.; Wright, Wallace; Hopkins, John A., Jr. *The Agricultural Emergency in Iowa.* Agricultural Experiment Station, Iowa State College of Agriculture and Mechanic Arts, Circulars 140-148. Ames: 1933.

The C.I.O. and the Farmers, Extracts from Speeches made at the First National Convention of the United Cannery, Agricultural, Packing and Allied Workers' Union of America. Denver: 1937.

Harter, William L. and Stewart, R. E. *The Population of Iowa: Its Composition and Changes, A Brief Sociological Study of Iowa's Human Assets.* Agricultural Experiment Station, Iowa State College of Agriculture and Mechanic Arts, Bulletin 275. Ames: 1930.

Sioux City Milk Producers Cooperative Association. *Record of Progress.* Sioux City: n.d.

Soth, Lauren K. *Agricultural Economic Facts Basebook of Iowa.* Agricultural Economics Subsection, Rural Social Science and Economics Section, Agricultural Experiment Station and Extension Service, Iowa State College of Agriculture and Mechanic Arts, Special Report No. 1. Ames: 1936.

Books

Barlow, Lester. *What Would Lincoln Do?* Stamford, Conn.: The Non-Partisan League Publishing Co., 1931.

Bell, Daniel (ed.). *The New American Right.* New York: Criterion Books, 1955.

Bendix, Reinhard and Lipset, Seymour Martin (eds.). *Class, Status and Power.* Glencoe, Illinois: The Free Press [c1953], 1956.

Benedict, Murray R. *Farm Policies of the United States, 1790-1950: A*

Study of Their Origins and Development. New York: Twentieth Century Fund, 1953.

Benedict, Murray R. and Stine, Oscar C. *The Agricultural Commodity Programs.* New York: Twentieth Century Fund, 1956.

Bloor, Ella Reeve. *We Are Many.* New York: International Publishers, 1940.

Blum, John M. *From the Morgenthau Diaries: Years of Crisis, 1928-1938.* Boston: Houghton Mifflin, 1959.

Campbell, Angus *et al. The American Voter.* New York: Wiley, 1960.

Campbell, Christiana McFadyen. *The Farm Bureau and the New Deal.* Urbana: University of Illinois Press, 1962.

Chambers, Whittaker. *Witness.* New York: Random House, 1952.

Commager, Henry Steele (ed.). *Documents of American History* (7th ed.). New York: Appleton-Century-Crofts, 1963.

Drake, St. Clair and Cayton, Horace R. *Black Metropolis: A Study of Negro Life in a Northern City* (rev. ed.). New York: Harper Torchbooks, 1962.

Draper, Theodore. *American Communism and Soviet Russia.* New York: Viking, 1960.

Federal Theatre Project. *Federal Theatre Plays: 1. Triple-A Plowed Under, by the Staff of the Living Newspaper, 2. Power, A Living Newspaper by Arthur Arent, 3. Spirochete, A History.* New York: Random House, 1938.

Fite, Gilbert C. *George N. Peek and the Fight for Farm Parity.* Norman: University of Oklahoma Press, 1954.

Fitzgerald, D. A. *Corn and Hogs Under the Agricultural Adjustment Act.* Washington: The Brookings Institution, 1934.

———. *Livestock Under the A.A.A.* Washington: The Brookings Institution, 1935.

Foster, William Z. *History of the Communist Party of the United States.* New York: International Publishers, 1952.

Glazer, Nathan. *The Social Basis of American Communism.* New York: Harcourt, Brace, 1961.

Hofstadter, Richard. *The Age of Reform.* New York: Alfred A. Knopf, 1955.

———. *The American Political Tradition.* New York: Vintage Books, 1954.

Howe, Irving and Coser, Lewis. *The Communist Party: A Critical History.* New York: Frederick A. Praeger, Inc., 1962.

Jones, Lawrence A. and Durand, David. *Mortgage Lending Experience in Agriculture.* A study by the National Bureau of Economic Research. Princeton: Princeton University Press, 1954.

King, Wendell. *Social Movements in the United States.* New York: Random House, 1956.

Kramer, Dale. *The Wild Jackasses: The American Farmer in Revolt.* New York: Hastings House, 1956.

Lipset, S. M. *Agrarian Socialism: The Cooperative Commonwealth Federation in Saskatchewan*. Berkeley: University of California Press, 1950.

Loomis, Charles P. and Beegle, J. Allan. *Rural Social Systems*. New York: Prentice-Hall, Inc., 1950.

Lord, Russell. *The Wallaces of Iowa*. Boston: Houghton Mifflin Company, 1947.

McCoy, Donald R. *Angry Voices: Left of Center Politics in the New Deal Era*. Lawrence: University of Kansas Press, 1958.

McCune, Wesley. *The Farm Bloc*. Garden City, N.Y.: Doubleday, Doran and Co., 1943.

Mayer, George H. *The Political Career of Floyd B. Olson*. Minneapolis: University of Minnesota Press, 1951.

Neprash, Jerry Alvin. *The Brookhart Campaigns in Iowa, 1920-1926*. New York: Columbia University Press, 1932.

Pollack, Norman. *The Populist Response to Industrial America*. Cambridge: Harvard University Press, 1962.

Rice, Stuart A. *Farmers and Workers in American Politics*. New York: Columbia University Press, 1924.

Robinson, Edgar Eugene. *The Presidential Vote, 1896-1932*. Stanford: Stanford University Press, 1934.

———. *They Voted for Roosevelt: The Presidential Vote, 1932-1944*. Stanford: Stanford University Press, 1947.

Robertson, Ross M. *History of the American Economy*. New York: Harcourt-Brace, 1955.

Roosevelt, Elliott (ed.). *F.D.R., His Personal Letters, 1928-1945*. 4 vols. New York: Duell, Sloan and Pierce, 1947-50.

Rosenman, Samuel I. (comp.). *The Public Papers and Addresses of Franklin D. Roosevelt*. 13 vols. New York: Random House, 1938-50.

Saloutos, Theodore and Hicks, John D. *Agricultural Discontent in the Middle West, 1900-1939*. Madison: University of Wisconsin Press, 1951.

Schlesinger, Arthur M., Jr. *The Coming of the New Deal*. Boston: Houghton Mifflin, 1959.

———. *The Crisis of the Old Order*. Boston: Houghton Mifflin, 1957.

———. *The Politics of Upheaval*. Boston: Houghton Mifflin, 1960.

Sheldon, Addison E. *Land Systems and Land Policies in Nebraska*. Publications of the Nebraska State Historical Society, XXII. Lincoln: 1936.

Short, Mrs. Wallace. *Just One American*. Sioux City: privately printed, 1943.

Sorokin, Pitirim. *Social and Cultural Dynamics* (rev. and abr. in one vol.). Boston: Extending Horizons Books, 1957.

Wallace, Henry A. *New Frontiers*. New York: Reynal & Hitchcock, 1934.

White, Roland A. *Milo Reno: Farmers' Union Pioneer.* Iowa City: Athens Press, 1941.

Articles

Anstrom, George. "Class Composition of the Farmers' Second National Conference, Chicago, 1933," *The Communist,* XIII (January, 1934), 47-52.

[Archie, David E.]. "Times of Trouble: The Cow War," *The Iowan* (Shenandoah, Iowa), VII (April-May, 1959), 28-35, 52-53.

Barnett, John. "The United Farmers League Convention," *The Communist,* XIII (August, 1934), 810-19.

Belknap, George M. "A Method for Analyzing Legislative Behavior," *Midwest Journal of Political Science,* II (November, 1958), 377-402.

Blackorby, Edward C. "William Lemke: Agrarian Radical and Union Party Presidential Candidate," *Mississippi Valley Historical Review,* XLIX (June, 1962), 67-84.

Bliven, Bruce. "Milo Reno and His Farmers," *New Republic,* LXXVII (November 29, 1933), 63-65.

Browder, Earl. "Report of the Political Committee to the Twelfth Central Committee Plenum . . . , November 22, 1930," *The Communist,* X (January, 1931), 7-32.

Carey, James C. "The Farmers' Independence Council of America, 1935-1938," *Agricultural History,* XXXV (April, 1961), 70-77.

Case, H. C. M. "Farm Debt Adjustment During the Early 1930s," *Agricultural History,* XXXIV (October, 1960), 173-81.

Dahl, Leif. "Nebraska Farmers in Action," *New Republic,* LXXIII (January 18, 1933), 265-66.

Dalrymple, Dana G. "The American Tractor Comes to Soviet Agriculture: The Transfer of a Technology," *Technology and Culture,* V (Spring, 1964), 191-214.

Davenport, Walter. "Get Away from Those Cows," *Colliers,* LXXXIV (February 27, 1932), 10-11.

Dileva, Frank D. "Attempt to Hang Iowa Judge," *Annals of Iowa,* XXXII (July, 1954), 337-64.

———. "Frantic Farmers Fight Law," *ibid.* (October, 1953), 81-109.

———. "Iowa Farm Price Revolt," *ibid.* (January, 1954), 171-202.

Dodd, James W. "Resolutions, Programs and Policies of the North Dakota Farmers' Holiday Association, 1932-1937," *North Dakota History,* XXVIII (April-July, 1961), 107-17.

Edwards, E. Everett, "American Agriculture—the First 300 Years," in U.S. Department of Agriculture, *Farmers in a Changing World: Yearbook of Agriculture, 1940.* Washington: Government Printing Office, 1940, 171-276.

Feder, Ernest. "Farm Debt Adjustments During the Depression—the Other Side of the Coin," *Agricultural History,* XXXV (April, 1961), 78-81.

Ferkiss, Victor C. "Populist Influences on American Fascism," *Western Political Quarterly,* X (June, 1957), 350-73.

Fite, Gilbert C. "Farmer Opinion and the Agricultural Adjustment Act, 1933," *Mississippi Valley Historical Review,* XLVIII (March, 1962), 656-73.

———. "John Simpson: The Southwest's Militant Farm Leader," *ibid.,* XXXV (March, 1949), 563-84.

———, ed. "Some John A. Simpson–Franklin D. Roosevelt Letters on the Agricultural Situation," *Chronicles of Oklahoma,* XXVI (Autumn, 1948), 336-45.

George, Harrison. "Causes and Meaning of the Farmers' Strike and Our Tasks as Communists," *The Communist,* XI (October, 1932), 918-31.

Glass, J. Remley. "Gentlemen, the Corn Belt!" *Harpers,* CLXVII (July, 1933), 199-209.

Harris, Lement. "The Spirit of Revolt," *Current History,* XXXVIII (July, 1933), 424-29.

["Harrow"]. "The Factory Farm: A Discussion Article on the Party and the Farm Problem," *The Communist,* VII (December, 1928), 761-69.

Herbst, Josephine. "Feet in the Grass-roots," *Scribner's,* XCIII (January, 1933), 46-51.

Hoglund, A. William. "Wisconsin Dairy Farmers on Strike," *Agricultural History,* XXXV (January, 1961), 24-34.

Lenin, V. I. "Capitalism and Agriculture in America," *The Communist,* VIII (June, 1929), 313-18; (July, 1929), 395-401; (August, 1929), 473-77.

Loomis, Charles P. and Beegle, J. Allan. "The Spread of German Nazism in Rural Areas," *American Sociological Review,* XI (December, 1946), 724-34.

Lovett, Robert Morss. "Hue [sic] Long Invades the Midwest," *New Republic,* LXXXIII (May 15, 1935), 10-12.

McElveen, Jackson V. "Farm Number, Farm Size and Farm Income," *Journal of Farm Economics,* XLV (February, 1963), 1-12.

Murphy, Donald R. "The Farmers Go On Strike," *New Republic,* LXXII (August 31, 1932), 66-67.

Nichols, Jeannette P. "Silver Inflation and the Senate in 1933," *The Social Studies,* XXXV (January, 1934), 12-18.

Nixon, Herman C. "The Economic Basis of the Populist Movement in Iowa," *Iowa Journal of History and Politics,* XXI (July, 1933), 373-96.

Puro, H[enry]. "The Class Struggle in the American Countryside," *The Communist,* XII (June, 1933), 547-58.

———. "The Farmers Are Getting Ready for Revolutionary Struggles," *ibid.,* XIII (June, 1934), 569-80.

———. "The Tasks of Our Party in the Work Among the Farmers," *ibid.,* XII (September, 1933), 875-87.

Richman, A. B. "The Economics of American Agriculture," *The Communist,* VIII (January-February, 1929), 88-94.

Ruggles, Clyde O. "The Economic Basis of the Greenback Movement in Iowa and Wisconsin," *Proceedings of the Mississippi Valley Historical Association for the Year 1912-1913,* VI (1913), 142-65.

Saloutos, Theodore. "William A. Hirth, Middle Western Agrarian," *Mississippi Valley Historical Review,* XXXV (September, 1951), 215-32.

Shover, John L. "The Farm Holiday Movement in Nebraska," *Nebraska History,* XLIII (March, 1962), 53-78.

———. "Populism in the Nineteen-Thirties: The Battle for the AAA," *Agricultural History,* XXXIX (January, 1965), 17-24.

Slichter, Gertrude Almy. "Franklin D. Roosevelt and the Farm Problem, 1929-1932," *Mississippi Valley Historical Review,* XLIII (September, 1956), 238-58.

Smith, Warren B. "Norman Baker—King of the Quacks," *The Iowan,* VII (December-January, 1958-59), 16-21, 55.

Tontz, Robert L. "Membership of General Farmers' Organizations, United States, 1874-1960," *Agricultural History,* XXXVIII (July, 1964), 143-56.

Trimble, William. "Historical Aspects of the Surplus Food Production of the United States, 1862-1902," *Annual Report of the American Historical Association for the Year, 1918.* Washington, 1921, I, 223-239.

Tucker, W. P. "Populism Up to Date: The Story of the Farmers' Union," *Agricultural History,* XXI (October, 1947), 198-208.

"U.S. Agriculture and Tasks of the Communist Party, U.S.A.: A Draft Program Proposed by the Agricultural Committee of the C.C.C. for General Discussion," *The Communist,* IX (February, 1930), 104-20; (March, 1930), 280-85; (April, 1930), 359-75.

Wakeley, Ray E. "How to Study the Effects of Direct Action Movements on Farm Organizations," *Social Forces,* XII (March, 1934), 380-85.

Weyl, Nathaniel. "I Was in a Communist Unit with Hiss," *U.S. News and World Report,* January 9, 1953, 22-40.

Wilcox, Benton H. "An Historical Definition of Northwestern Radicalism," *Mississippi Valley Historical Review,* XXVI (December, 1939), 377-94.

INDEX